VIRAL PATHOGENESIS
in DIAGRAMS

Hans-W. Ackermann, M.D.
Laurent Berthiaume, Ph.D.
Michel Tremblay, Ph.D.

VIRAL PATHOGENESIS in DIAGRAMS

CRC Press
Boca Raton   London   New York   Washington, D.C.

Library of Congress Cataloging-in-Publication Data

Ackermann, Hans-Wolfgang, 1936-
  Viral pathogenesis in diagrams / Hans-W. Ackermann, Laurent Berthiaume, Michel Tremblay.
      p.   cm.
  Includes bibliographical references and index.
  ISBN 0-8493-2207-3 (alk. paper)
    1. Virus diseases--Pathogenesis--Atlases. I. Berthiaume, Laurent. II. Tremblay,
  Michel. III. Title.

QR201.V55 A25 2000
616′.0194—dc21   00-062154

© 2001 by CRC Press LLC

No claim to original U.S. Government works
International Standard Book Number 0-8493-2207-3
Library of Congress Card Number 00-062154
Printed in the United States of America 1 2 3 4 5 6 7 8 9 0
Printed on acid-free paper

CONTENTS

ACKNOWLEDGMENTS

This book was produced with the collaboration of several major scientific publishers. Academic Press, through its New York and London Divisions, provided 35 diagrams. This large number reflects the outstanding role of Academic Press as the publisher of *Advances in Virus Research, Virology,* and *Seminars in Virology.* The American Society of Microbiology, publisher of *Virology,* and the *Journal of Virology,* contributed 24 diagrams. Elsevier Science, mainly through its periodical *Trends in Virology,* made a further major contribution of 22 diagrams. Lippincott Williams & Wilkins provided 41 diagrams from various books.

We wish to thank the following persons for permission to use their figures in this book: R. Ahmed, J.W. Alexander, A.M. Arvin, H.H. Balfour, J.E. Banatvala, S. Baron, D.G. Bausch, J.A. Bellanti, M.-G. Bergeron, L. Bianchi, H. Bielefeldt-Ohmann, R. Blacklaws, B.S. Blumberg, D.K. Bodkin, D.P. Bolognesi, J.S. Butel, B.W. Calnek, S. Chirkov, R. Colimon, R.A. Consigli, A.M. Crawford, P.C. Doherty, A.L.W.F. Eddleston, A.S. Fauci, F. Fenner, H. Feldmann, D.B. Fishbein, A. Flamand, M.J. Fraser, N.W. Fraser, I.L. Garner, G. Garrigue, G. Gillet, H.J. Ginsberg, F. Gonzalez-Scarano, B.S. Graham, S.M. Gray, D.E. Griffin, C. Grose, P.E. Gunton, H. Hengel, M.G. von Herrath, A.B. Hill, M. Hilleman, M. Hiyoshi, P. Höllsberg, J.H. Hoofnagle, E.A. Hoover, S. Howell, P.M. Howley, B.T. Huber, S.A. Huber, J.M. Huraux, R.T. Johnson, C. Jones, J.F.R. Kerr, S.C. Knight, D.L. Kolson, L.G. Koss, I. Kurane, S.P. Lee, P. Leinikki, S.M. Leisner, A.J. Levine, J.A. Levy, T. Magrath, K. Maletic Neuzil, V. Maréchal, G.A. Martini, G. McFadden, F. Miedema, T.P. Monath, L.A. Morrison, A.J. Nahmias, A.A. Nash, N. Nathanson, M.L. Nibert, A.L. Notkins, M.B.A. Oldstone, G. Orth, T.M. Paronson, N.C. Pedersen, S. Perlman, P. Piot, J.H.L. Playfair, R.H. Purcell, N. Raab-Traub, J. Rajcáni, F. Rapp, C.E. Rogler, I. Roitt, Z.F. Rosenberg, F. Rozenberg, D.H. Rubin, C.E. Samuel, S. Schneider-Schaulies, J. Schüpbach, G.C. Sen, D.A. Shafritz, A.H. Sharpe, S. Sherlock, K. Shimotohno, T.J. Smith, W.A. Stevens, W. Stille, S.E. Straus, G.A. Strobel, P.E. Tambourin, M. Tashiro, H.C. Thomas, D.A. Thorley-Lawson, G. Trinchieri, E.R. Unanue, M. Vasseur, L.E. Volkman, B.A. Webb, M.C. Weissenbacher, R.M. Welsh, E. White, R.J. Whitley, J.L. Whitton, T.C. Wright, K. Yamaguchi, D.L. Yirrell, M. Yoshida, N.S. Young, M. Zink, F. Zoulim, H. Zur Hausen. In particular, this book would not have come into being without the numerous and beautiful diagrams by C.A. Mims, A.S. Fauci, and R.T Johnson, which contribute much to the understanding of viral pathogenesis.

Our special thanks go to the ICTV secretary, Dr. C.R. Pringle, University of Warwick, Coventry, England, for making unpublished ICTV material available to us. Jeannine Gauthier, Quebec, carried out bibliographical research. Michel Côté, Rexfor, Quebec, skillfully prepared figures 24, 58, 162, 177, 179, 182, and 233. Suzanne Bernatchez and Denise Raby, Laval University, provided editorial help in the final stages of the manuscript.

THE AUTHORS

Hans-Wolfgang Ackermann, M.D., is Professor of Microbiology at the Medical School of Laval University, Quebec, Canada. He was born in Berlin, Germany, in 1936 and obtained his medical degree in 1962 at the Free University of Berlin (West). He was a fellow of the Airlift Memorial Foundation and received much of his microbiological training at the Pasteur Institute of Paris.

After a period of research and teaching at the Free University, he left Germany in 1967 and went to Canada, where he started to investigate bacteriophage morphology. During his career, he has done research on enterobacteria, airborne fungi, human hepatitis B virus, and baculoviruses; however, bacteriophages have always been the center of his interest. He teaches virology, electron microscopy, and mycology and has a strong interest in audiovisual teaching aids. In 1982, he founded the Félix d'Hérelle Reference Center for Bacterial Viruses, which is essentially a culture collection aimed at the preservation of type phages. Dr. Ackermann is author or co-author of about 150 scientific papers or book chapters and senior author of three books, entitled *Viruses of Prokaryotes, Atlas of Virus Diagrams,* and *Virus Life in Diagrams* (CRC Press, 1987, 1995, and 1998, respectively). He is a member of the International Committee on Taxonomy of Viruses (ICTV) since 1971 and several times was chairman or vice-chairman of its Bacterial Virus Subcommittee, was ICTV vice-president (1984-1990), and was a member of the Executive Committee until 1996. He is now a life member of the ICTV.

Laurent Berthiaume, Ph.D., is a former professor of the Virology Center of the National Institute for Scientific Research (INRS), a branch of the University of Quebec at Montreal. He was born in Montreal in 1941 and obtained his Ph.D. degree in microbiology-immunology in 1972 at the University of Montreal. He was trained in electron microscopy at the University of Toronto School of Hygiene, was in charge for many years of the electron microscopic laboratory of his institute, and eventually became coordinator of graduate studies and registrar.

During his activities as a virologist and electron microscopist, he developed a strong interest in virus morphology and replication and the rapid diagnosis of viral infections. During many years, he lectured on virus structure and taxonomy. His most recent work has centered on viruses infecting fish, especially on the antigenic and genetic diversity of aquabirnaviruses. He is co-author of the *Atlas of Virus Diagrams* and of *Virus Life in Diagrams* and the author of about 85 scientific papers or book chapters and 150 communications at congresses. He has been an ICTV member since 1975 and a member of the ICTV Executive Committee from 1989 to 1995. He is presently retired and a freelance writer in virology.

Michel J. Tremblay, Ph.D., is Professor of Microbiology at the Medical School of Laval University. He was born in Jonquière, P.Q., in 1955 and obtained his Ph.D. degree in experimental medicine in 1989 from McGill University. After post-doctoral work in molecular immunology at the Montreal Clinical Research Institute, he founded in 1991 a laboratory for human retrovirology at the Laval University Hospital. Dr. Tremblay is the author or co-author of 65 papers and over 130 communications at congresses, and co-author of *Virus Life in Diagrams.*

His research centers on immunological and virological aspects of the human immunodeficiency virus (HIV), the etiologic agent of AIDS. He is studying intracellular signal transduction after contact between the virus particle and its target cell, the modulatory role of tyrosine phosphatases on HIV transcription, putative bidirectional interactions between *Leishmania* and HIV, and the acquisition of host-encoded glycoproteins by nascent HIV and their role in the viral life cycle. Dr. Tremblay is a member of the International AIDS Society and the Canadian Association for HIV Research.

Chapter 1

INTRODUCTION

This book is a sequel to the *Atlas of Virus Diagrams,* published in 1995 and concerned with viral structure and morphology, and to *Virus Life in Diagrams,* published in 1998 and dedicated to morphogenesis and replication.[1,2] Its aim is to assemble the many diagrams on the pathogenesis of viral diseases that are scattered over the literature. As with its predecessors, it is aimed at teachers and students. It has an unusually wide scope.

Pathogenesis may be defined as the development of a disease. In virology, this term comprises entry of the virus into the host, its dissemination, humoral and cellular host defenses, self-destructive autoimmunity, evasion of defenses by the virus, and molecular mechanisms of cell death. For many authors, pathogenesis is more or less synonymous with symptomatology. Viral pathogenesis thus includes not only virology, but also much of clinical human and veterinary medicine, epidemiology, immunology, molecular biology, and pathology. For added complexity, this book includes insect and plant viruses.

Diagrams are extensively used in virology to illustrate morphology and morphogenesis, genome structure and replication, epidemiology, pathogenesis, and virological techniques. Their function is to summarize and integrate large numbers of observations, from electron microscopy to clinical data, into a single picture or a few related drawings. Diagrams illustrate processes and thereby transcend photographs and experimental data, which usually reflect the state or activity of a virus at a given moment. Diagrams are thus ideally suited to illustrate the pathogenesis of viral diseases from infection to host defenses and cell death.

Some diagrams are historical documents that illustrate the development of virology. For example, the progress of herpes simplex virus in the ophthalmic branch of the trigeminus nerve is illustrated in a diagram from 1932 (Fig. 117) and cellular effects of nuclear polyhedrosis and granulosis viruses of insects are demonstrated in drawings from the same time. The first, justly famous, diagram to illustrate the complete pathogenesis of a viral disease is from 1948 and shows the course of mousepox (Fig. 183). Other diagrams from the mid-fifties show that the course of poliomyelitis was well understood at that time. These and other older diagrams, though outdated, are reproduced as such. On the other hand, certain very recent diagrams are based on research in progress and contain speculative elements.

The 284 diagrams of this book are from over 40 periodicals and 70 books or monographs and have been selected from over 800 diagrams from the English and French virological literature. We found no suitable diagrams in other language domains. Many diagrams are from a few journals specializing in review papers, namely *Advances in Virology, Seminars in Virology,* and Trends in *Microbiology.* A few diagrams are our own. One of them is derived from a famous drawing by Leonardo da Vinci. We believe that this great man and universal scientist would not have been averse to the use of his drawing in a virological text.

Scientific content was the prime reason for selection, followed by clarity and didactic value, originality of concept, esthetic appeal, and historic interest. We did not favor simple text diagrams, consisting of lettering only and devoid of illustrative components.

A number of diagrams illustrate related subjects and may appear as duplications. In fact, they are complementary because there are often many ways to relate or explain the same observation. However, many interesting and valuable diagrams were omitted for lack of space. Other reasons for omission were speculative content, difficulty of reproduction, e.g., of color diagrams, and prohibitive fees. Indeed, it is now commonplace for North American publishers to ask for high fees and this may well be the last time that a collection of diagrams is attempted. It is apparently not realized that the authors of a book such as the present one do not make any profits and that their sole motivation is the progress of science.

An overview of virus classification is presented in Chapter 2. Viruses are grouped by nature of nucleic acid, presence or absence of an envelope, and capsid symmetry. The reader is referred to publications of the International Committee on Taxonomy of Viruses (ICTV) for a detailed description of taxa. An introduction to vertebrate virus pathogenesis is found in Chapter 3. The similarly general Chapters 4 and 5 cover types of disease, spread of viruses through the body, cytopathology, and host defenses. Chapter 6 illustrates the behavior of 19 vertebrate virus families in alphabetical order. They include viruses replicating in vertebrates, arthropods, or plants ("arboviruses"), e.g., bunyaviruses, reoviruses, and togaviruses. The nonassigned deltaviruses are added to the *Hepadnaviridae* family because of their dependence on hepadnaviruses and similar epidemiology. Virus families are briefly characterized by key words indicating basic features (nature of DNA or RNA, capsid symmetry, presence or absence of an envelope) and short paragraphs describing taxonomical status, particle structure, host range, physiological properties, and salient features of specific viral diseases caused by their members. In a general way, vertebrate viruses cannot be divided into major pathogenetic groups according to particle structure or nature of nucleic acid. Chapters 4 to 6 are limited to warm-blooded animals because diagrams relating to lower vertebrates seem to be nonexistent.

Chapters 7 and 8 comprise introductions to the pathogenesis of insect and plant viruses, respectively. Insect virus diagrams are essentially limited to baculoviruses. Diagrams on plant viruses illustrate infection, virus spread, and cytopathology rather than specific diseases. The effect of bacterial viruses on bacterial cells (lysis, lysogenization) has been illustrated in *Virus Life in Diagrams* and is not considered here. To our knowledge, the literature contains virtually no illustrations of virus infections of fungi, algae, protozoa, and other invertebrates than insects.

Diagrams are thus grouped into five chapters: (4 and 5) general pathogenesis in vertebrate virus infections, (6) vertebrate viruses by family or group, (7) insect viruses, and (8) plant viruses. Over 80% of available diagrams are on vertebrate viruses. Only less than 5% and 15% of diagrams illustrate insect and plant viruses, respectively.

In families illustrated more than once, diagrams are arranged from the general to the detailed, e.g., starting with the whole organism or cell and ending with molecular mechanisms of pathogenesis. The information content reflects current research interests or funding and is very uneven. For example, adenovirus diagrams focus on apoptosis only. On the other hand, there is a wealth of diagrams on AIDS and herpesvirus pathogenesis down to intricate and often highly specialized and speculative diagrams of molecular pathogenesis. In a general way, the present understanding of the molecular pathogenesis of viral diseases is most uneven and limited. Each virus group seems to elicit its own specific set of cytokines and host responses. No general pattern is yet apparent and the story of the molecular basis of viral diseases remains to be written. This is especially true for plant virus diseases, in which our understanding of pathogenesis is essentially limited to virus movement through plants.

The diagrams were recorded with a SuperVista S-12 high-resolution scanner (UMAX Technologies, Freemont, CA; 600 dpi). Damaged diagrams, for example from microfilms, were restored and French-language captions and legends were translated into English. Color diagrams were rendered in black and white. Legends were generally reproduced without changes, but some, extending over entire pages, were extremely long and had to be shortened. Others, very short and meaningless when taken out of context from the original publication, were completed. We considered each diagram as an independent unit that had to be understandable in itself. Faulty English was corrected on occasion.

Due to the wide scope of the subject of pathogenesis and its rather imprecise boundaries, there is a profusion of abbreviations and scientific terms. Abbreviations, particularly frequent in immunology, are listed in an appendix; however, the latter excludes abbreviations explained in the text and used only once. Scientific terms are explained in an extensive glossary. Terms from human medicine are considered as widely known and are generally not explained, except when they include personal names (e.g., Koplik's spots or Kupffer cells), or must appear unclear to the nonmedical reader (e.g., carcinoma *in situ*). Definitions are generally taken from dictionaries of general and molecular biology, immunology, and virology;[3-6] a few have been devised by us. Many terms have different meanings in different disciplines. This is indicated by abbreviations such as BOT for botany and IMM for immunology, respectively.

Chapter 2

A SUMMARY OF VIRUS CLASSIFICATION

Viruses minimally consist of nucleic acid and a protein shell or capsid, The shell is of cubic or helical symmetry or, in tailed phages, a combination thereof ("binary symmetry"). Capsids with cubic symmetry are icosahedra or related bodies. In about one third of virus families, the capsid is surrounded by a lipid-containing envelope. A few exceptional types have an envelope and no capsid. The nucleic acid is single-stranded or double-stranded DNA or RNA, is linear or circular, and comprises one or several molecules (up to 12). All DNA viruses except polydnaviruses contain a single molecule of DNA. All DNA genomes except that of hepatitis deltavirus are linear. In some plant viruses, the individual parts of multipartite genomes are packaged into separate shells, constituting multicomponent systems. Bringing these basic disparate facts into a coherent and comprehensive system is the task of the International Committee on Taxonomy of Viruses or ICTV.

The ICTV developed from the International Committee on Nomenclature of Viruses (ICNV). It was established in 1963, became permanent in 1966 at the XIth International Congress of Microbiology in Moscow, and was renamed ICTV in 1973.[7] The ICTV consists of 6 subcommittees, about 45 study groups, and over 400 participating virologists. It has subcommittees for viruses of vertebrates, invertebrates, plants, bacteria, eukaryotic protists (algae, fungi, protozoa), and the creation of a viral database. Study groups are formed by co-optation as the need arises. Taxonomic and nomenclature proposals usually originate in study groups and are successively voted on by the respective subcommittees, the ICTV Executive Committee, and the full ICTV when it convenes every three years at an International Congress of Virology. Only then do they become official. Taxa are subject to constant revision.

The ICTV is concerned with a universal coherent system of virus classification and nomenclature and issues periodically a report, in principle after each International Congress of Virology. The report describes virus orders, families, subfamilies, and genera. Species are simply listed without further details. The last two reports contain numerous micrographs, genomic maps, evolutionary trees, and diagrams illustrating replication strategies.[8]

In principle, viruses are classified with the help of all available data. Some criteria are of particular value, namely the presence of DNA or RNA, type of capsid symmetry, and presence or absence of an envelope. These criteria were introduced by Lwoff, Horne, and Tournier[9] as the basis of a hierarchical system of viruses. The system did not survive, but its major criteria stood the test of time. More high-level criteria were added later. They relate to the genome (strandedness, message-sense [+] or anti-message [-] strands,

number of segments) and replication (e.g., presence of a DNA step or RNA polymerase). Most criteria are virus-dependent. Host-related criteria such as symptoms of disease are few and of limited importance. Genome structure and base or amino acid alignments are becoming increasingly important as data become available. The ultimate goal is a phylogenetic system of viruses. Viruses are essentially classified by the following criteria:

1. Nucleic acid: DNA or RNA, number of strands; conformation (linear, circular, superhelical), sense (+ or -), nucleotide sequence, presence or absence of 5'-terminal caps, terminal proteins, or poly(A)tracts.
2. Morphology: size, shape, presence or absence of an envelope or peplomers, capsid size and structure.
3. Physicochemical properties: particle mass, buoyant density, sedimentation velocity, and stability (pH, heat, solvents, detergents).
4. Proteins: content, number, size, and function of structural and nonstructural proteins, amino acid sequence, glycosylation.
5. Lipids and carbohydrates: content and nature.
6. Genome organization and replication: gene number and genomic map, characteristics of replication, transcription and translation, post-translational control, site of particle assembly, mode of release.
7. Antigenic properties.
8. Biological properties: host range, transmission, geographic distribution, cell and tissue tropism, pathogenicity and pathology.

The present system of viruses includes 3 orders, 62 families, and 238 genera.[8, 10] In a departure from former stances, the ICTV has now classified retrotransposons, e.g., the copia agent of *Drosophila* and yeast Ty1 and Ty3 agents (*Metaviridae* and *Pseudoviridae*), and capsidless replicating RNA (*Narnaviridae*). The ICTV also records, without attempts at classification, viroids, satellite RNAs, and prions. These agents are on the fringe, or outside, of the viral world and are not reported in Table 1.

The order *Mononegavirales*, characterized by the presence of a single molecule of negative-sense ssRNA, comprises the *Filoviridae, Paramyxoviridae,* and *Rhabdoviridae* families. The order *Nidovirales* includes the *Arteriviridae* and *Coronaviridae*. The recently created order *Caudovirales* comprises the vast group of tailed bacteriophages.[8,11] More viral orders may be established in the future. Based on genome organization and amino acid sequence similarity, essentially of RNA polymerases and helicases, three "superfamilies" have been individualized among ssRNA viruses, namely picorna-like, Sindbis-like, and luteovirus-like viruses.[12] These groups are fairly heterogeneous and comprise animal and plant viruses, rod-shaped, icosahedral, enveloped and nonenveloped particles, and viruses with monopartite or multipartite genomes. It is yet unclear to which extent these similarities, involving a few genes in small genomes, are due to horizontal gene transfer.

Families and genera are the most stable parts and the backbone of the system. About 30 "floating" genera, mostly of plant viruses, have yet to be assigned to a higher taxon. Although the *Poxviridae* family has 11 genera, many other families are monogeneric and some even consist of a single member. The ICTV has also adopted the polythetic species concept, meaning that a species is defined as a virus by a set of properties, all of which are not necessarily present in all members. A virus species is defined as a "polythetic class of viruses that constitutes a replicating lineage and occupies a particular ecological niche".[13]

Orders, family, subfamily, and genus names end by the Latin suffixes *-virales, -viridae, -virinae,* and *-virus,* respectively. Names always reflect virus characteristics. The creation of latinized binomial species names is not attempted. Table 1 summarizes the state of virus classification. It contains the taxa adopted at the XIth International Congress of Virology in Sydney (1999).[8,10] Virus taxa are listed by nature of nucleic acids, presence or absence of an envelope, and capsid symmetry. While most viruses have cubic or helical nucleocapsids, a few types (*Fuselloviridae, Plasmaviridae,* the genus *Umbravirus*) have an envelope only and cannot be attributed with certainty to any symmetry class.

Table 1
VIRUS FAMILIES AND NONCLASSIFIED GENERA

Genome	Envelope	Capsid Symmetry	Order or Family	Non-classified	Number of Genera	Host
ssDNA	No	Cubic	*Circoviridae*		2	V, P
			Geminiviridae		4	P
			Microviridae		4	B
			Parvoviridae		6	V, I
			-	*Nanovirus*	1	P
	No	Helical	*Inoviridae*		2	B
dsDNA	Yes	Cubic	*Asfarviridae*		1	V
			Herpesviridae		10	V
	Yes	Helical	*Ascoviridae*		1	I
			Baculoviridae		2	I
			Lipothrixviridae		1	B
			Polydnaviridae		2	I
			Poxviridae		11	V, I
	Yes	None	*Fuselloviridae*		1	B
			Plasmaviridae		1	B
	No	Cubic	*Adenoviridae*		2	V
			Corticoviridae		1	B
			Iridoviridae		4	V, I
			Papillomaviridae		1	V
			Phycodnaviridae		4	A
			Polyomaviridae		1	V
			Tectiviridae		1	B
			-	*Rhizidiovirus*	1	V
		Helical	*Rudiviridae*		1	B
		None	-	*Sulfolobus-SNDV*	1	B
			Caudovirales:			
	No	Binary	" *Myoviridae*		6	B
			" *Siphoviridae*		6	B
			" *Podoviridae*		3	B
dsDNA, reverse	Yes	Cubic	*Hepadnaviridae*		2	V
	No	Cubic	*Caulimoviridae*		6	P
ssRNA, reverse	Yes	Helical	*Retroviridae*		7	V
ssRNA, (+) sense	Yes	Cubic	*Flaviviridae*		3	V, I
			Togaviridae		3	V, I
	Yes	Helical	**Nidovirales:**			
			" *Arteriviridae*		1	V
			" *Coronaviridae*		2	V
	Yes	?		*Umbravirus*	1	P
	No	Cubic	*Astroviridae*		1	V
			Bromoviridae		5	P
			Caliciviridae		4	V
			Comoviridae		3	P
			Leviviridae		2	B
			Luteoviridae		3	P

VIRUS FAMILIES AND NONCLASSIFIED GENERA (continued)

Genome	Envelope	Capsid Symmetry	Order or Family	Non-classified	Number of Genera	Host
			Nodaviridae		2	I
			Picornaviridae		9	V
			Sequiviridae		2	P
			Tetraviridae		2	I
			Tombusviridae		8	P
			-	Hepatitis E-like	1	V
			-	Cricket paralysis	1	I
			-	*Idaeovirus*	1	P
			-	*Marafivirus*	1	P
			-	*Sobemovirus*	1	P
			-	*Tymovirus*	1	P
	No	Helical rods	*Barnaviridae*		1	F
			-	*Benyvirus*	1	P
			-	*Furovirus*	1	P
			-	*Hordeivirus*	1	P
			-	*Pecluvirus*	1	P
			-	*Pomovirus*	1	P
			-	*Tobamovirus*	1	P
			-	*Tobravirus*	1	P
			-	*Ourmiavirus*	1	P
	No	Helical filaments	*Closteroviridae*		2	P
			Potyviridae		6	P
			-	*Allexivirus*	1	P
			-	*Capillovirus*	1	P
			-	*Carlavirus*	1	P
			-	*Foveavirus*	1	P
			-	*Potexvirus*	1	P
			-	*Trichovirus*	1	P
			-	*Vitivirus*	1	P
(-) sense, monopartite	Yes	Helical	**Mononegavirales:**			
			" *Bornaviridae*		1	V
			" *Filoviridae*		2	V
			" *Paramyxoviridae*		5	V
			" *Rhabdoviridae*		6	V, I, P
	?		-	*Deltavirus*	1	V
(-) sense, multipartite	Yes	Helical	*Arenaviridae*		1	V
			Bunyaviridae		5	V, I, P
			Orthomyxoviridae		4	V
	No	Helical	-	*Ophiovirus*	1	P
			-	*Tenuivirus*	1	P
dsRNA	Yes	Cubic	*Cystoviridae*		1	B
		?	*Hypoviridae*		1	F
	No	Cubic	*Birnaviridae*		3	V, I
			Partitiviridae		4	F, P
			Reoviridae		9	V, I, P
			Totiviridae		3	F, Pr
		Helical		*Varicosavirus*	1	P

A, algae; B, bacteria; F, fungi; I, invertebrates; P, plants; Pr, protozoa; V, vertebrates.

Chapter 3

VIRAL PATHOGENESIS IN VERTEBRATES: PRINCIPLES

Vertebrate viruses belong to 26 families and 84 genera, including the unassigned genus *Deltavirus* (D hepatitis agent). Virions are enveloped or not and contain single-stranded or double-stranded DNA or RNA. Large viruses with dsDNA and complex morphology are relatively frequent and viruses with single-stranded (+) RNA are relatively rare. Viruses of the *Retroviridae* family contain ssRNA and use reverse transcription from RNA to DNA for replication. A few dsDNA viruses, called "pararetroviruses", practice reverse transcription from DNA to RNA. Certain viruses, notably members of the *Bunya-, Reo-, Rhabdo-,* and *Togaviridae* families, replicate in both vertebrates and arthropods (insects, ticks) and are called arboviruses.

Viral pathogenesis can be defined as the capacity of a virus to cause disease in the host. The initial stage in viral pathogenesis is entry into the host which can be mediated by direct inoculation through the skin (for example by bites of arthropod vectors) or via the mucosal surfaces (i.e., respiratory, gastrointestinal, and genito-urinary tract, conjunctiva). Once inside the host, the initial replicative cycle of the virus takes place in a primary site generally located near the portal of entry. Spread through the host is achieved either by cell-free viruses or passage from cell to cell. Viral pathogenesis is a highly complex process and its nature is commonly multifactorial since more than one single factor is leading to a disease. It is important to state that virulence, which is the ability of a virus to produce a disease, depends on a variety of both viral (e.g., size of inoculum, route of entry) and host factors (e.g., age, species, and immunocompetence). A dynamic balance between the host and the virus dictates the course of a virus infection. Viruses have evolved a wide array of mechanisms for mediating pathologic effects in the infected host such as direct cell damage resulting from virus replication (e.g., apoptosis, anergy, lysis, syncytium formation, alteration of signal transduction events, cell transformation) and viral effects on host immune response (e.g., immunosuppression, immune activation). *In vivo* experimentation aimed at studying viral pathogenesis seen in humans is technically and ethically difficult. Many investigators have used animal models to acquire insights into how viruses produce human diseases because such systems can be more readily manipulated and studied.

The interaction between a virus and a host may lead to a variety of outcomes, including acute, persistent, latent, slow virus, or silent infection and induction of tumors.

1. Acute infection is seen in several classical virus diseases, such as measles, smallpox, rabies, and influenza.

2. Persistent infections are those in which the virus can be continuously detected and mild or no clinical symptoms are present.
3. Latent infections are those in which the virus persists in a cryptic form most of the time and intermittent symptoms may be diagnosed on occasion.
4. Slow virus infections are characterized by a prolonged incubation period during which virus replication is still occurring. Clinical symptoms are usually present during the incubation period.
5. Silent infections produce no apparent clinical signs.
6. Oncogenesis is a complex, multi-step process, resulting from changes in the morphological, biochemical, or growth parameters of a cell.

The ability of a virus to persist in the host often determines the outcome of a virus infection. Long-term persistence of viruses results from two main mechanisms. The first is the regulation of viral lytic potential which can be accomplished by restricting virus gene expression (e.g., herpesviruses). The second mechanism is evasion of immune surveillance by using strategies such as integration of the viral genome or adoption of an episomal form, infection of immunologically inaccessible anatomic sites, production of antigenic variants, interference with T-cell recognition, and induction of tolerance.

Many factors that determine the degree of illness are still undefined. The production of a disease is an abnormal and fairly unusual event in viral infections because the vast majority of virus infections are silent or subclinical (i.e., asymptomatic). Evidently, viruses tend to injure as less as possible their hosts. The capacity of a virus to mediate outward signs of disease is frequently linked to mechanisms adopted by viruses for avoiding host antiviral defenses (i.e., the immune response).

A number of morphological changes can be detected by microscopic observation of virus-infected cells. They arise from the generation of viral proteins and nucleic acids, but also from disruption of the biosynthetic capabilities of the host cell. Viruses accomplish this task by sequestering cellular organelles such as ribosomes[14] which are normally devoted to the synthesis of proteins essential to normal functions of the cell. Infection of cells with viruses induces several phenotypic changes that are often referred to as the cytopathic effects of a virus. The most common morphological modifications include altered shape, detachment from the substrate, lysis, membrane fusion (also termed syncytium formation), and production of inclusionbodies. Lysis represents the most extreme example of phenotypic changes in virus-infected cells. In this case membrane integrity is lost and the entire cell breaks down. Not all viruses induce such a dramatic effect. Lysis is advantageous to a virus because it favors the release of newly formed viral entities into the surrounding environment. Viruses causing cell lysis also frequently mediate a sudden and complete cessation of most host cell macromolecular synthesis, a process better known as shut-off. A clear example of shut-off is provided by the poliovirus 2 protein, a viral protease which inhibits the formation of a functional elongation factor necessary for cap-dependent translation of messenger RNAs by ribosomes.[14] Syncytium formation results from membranes of adjacent cells fusing together through the action of virus-encoded glycoproteins from enveloped viruses. Membrane fusion is a common feature of infection with Sendai virus, herpes simplex virus, and human immunodeficiency virus (HIV).

Viruses may also trigger in infected target cells a pathway of programmed cell death (apoptosis), which is characterized by distinct morphological changes such as the formation of condensed fragments of nuclear chromatin. Apoptosis is an energy-dependent process that results from activation of a calcium-dependent endonuclease whose function is to cleave cellular genomic DNA at each internucleosomal bridge. The phenomenon of apoptosis seems to play an important role in the control of virus infection since several viruses have developed ways to counteract apoptosis.[15] Herpes virus, Epstein-Barr virus, adenoviruses, and African swine fever virus have all been reported to code for proteins that function as inhibitors of apoptosis.

A main feature of several viral infections is the induction of an immunosuppressive state very early after infection. This virus-induced immunosuppression is characterized by a diminished ability to proliferate in response to several stimuli. The cellular basis for the non-responsiveness is still not fully understood although various studies conducted with HIV have shown that some viral proteins display immunosuppressive effects.

Cellular transformation may result in changes in morphological, biochemical, or growth parameters that can be detected in a cell infected with viruses. Transformed cells do not automatically lead to the development of tumors in experimental animals. Transformed cells exhibit a modified phenotype, which is displayed by a loss of anchorage dependence, loss of contact inhibition, colony formation in semi-solid media, and decreased requirement for growth factors. Proteins encoded by retroviral oncogenes mediate cell transformation. Some oncogenes are interfering with the normal process of signal transduction while others are implicated in the control of the cell cycle. The transforming genes of DNA tumor viruses have no cellular counterparts, as it is the case for the oncogenes of retroviruses. They achieve their effects by interacting with nuclear proteins involved in the control of DNA replication. The relationship between virus infection and tumorigenesis has been established for human T-cell lymphotropic virus type I (HTLV-I) in leukemia, Epstein-Barr virus in Burkitt's lymphoma and nasopharyngeal carcinoma, and hepatitis B virus in hepatocellular carcinoma.

It is now clear that the final outcome of viral infection in a given host depends not only on viruses, but also on a number of host factors. Experiments conducted with animal models clearly indicate that the genetic background of the host represents one of the most crucial factors influencing the outcome of a viral infection. Genetic factors modulating susceptibility to infection of certain strains of mice have been revealed following infection with flaviviruses (e.g., St. Louis encephalitis virus, yellow fever virus, West Nile virus), influenza viruses, arena-, corona-, herpes-, orthomyxo-, papova-, pox-, rhabdo-, and retroviruses, and with the agent of Borna disease. A strong correlation between the age of the host and the severity of viral infection has also been established when a large human population is exposed to the same viral pathogen. For example, a more severe disease is often produced in newborn animals. The exact reason of this age dependence is still poorly understood, but it seems clear that some types of inherited age-related resistance to viral infection may depend on the maturation or decline of both specific and nonspecific components of the host's immune response. The pathogenesis of a viral infection can also be influenced by the propensity of some viruses for preferential replication in actively dividing cells, which can quantitatively differ depending on the organ and the maturation stage of the target cell. The outcome of viral infection can be modulated at some extent by hormones. This is illustrated by the observation that male mice are more susceptible to some specific viruses than their female counterparts. Viral pathogenesis is also markedly influenced by the nutritional state of the host. The best example of this is the finding that protein malnutrition dramatically amplifies the severity of measles, coxsackievirus, and flavivirus infections.

The immune system appears as the controlling factor within the host that maintains beneficial microbes at harmless levels and prevents infection by dangerous agents. The immune system of the host is able to combat a variety of infections from birth on. This is accomplished by a system of barriers conferring a generalized or innate immunity. It comprises physical barriers to microbial entry, specific phagocytic cells (e.g., macrophages), eosinophils, basophils, natural killer cells, and various soluble factors, notably the "interferon" complex discovered in the fifties.[16] Interferons are induced upon infection of a variety of cells with viruses. These proteins can trigger the synthesis of several host-cell proteins that contribute to the inhibition of viral replication (e.g., 2'-5'-oligo-adenylate synthetase), activate a serine/threonine kinase called P1 kinase, increase expression of the MHC-I and TAP transporter proteins, and, finally, activate NK cells. The host also possesses an adaptive, specific immunity constituted of humoral and cellular elements, mediated by B cells and antibodies and by T cells, respectively. T cells can recognize foreign antigens as peptides bound to proteins of the major histocompatibility complex class I and II (MHC-I and MHC-II) molecules. Innate immunity is present at all times while adaptive immunity is induced by antigens and gives rise to a long-lasting protection against disease.

Since viral pathogenesis is markedly influenced by host response to viral infection, it is not surprising that disease symptoms are sometimes attributable not to viral replication per se, but to side-effects of the immune response. For example, cells of the immune system may release upon virus infection a series of potent soluble factors (e.g., interleukins, chemokines, and interferons) responsible for the apparition of symptoms such as inflammation, fever, headaches, and skin rashes. Evolutionary pressure has forced some viruses to develop numerous strategies to subvert the immune system. A very commonly used strategy employed by viruses is the production of antigenic variants. This strategy is adopted by influenza virus and HIV to inhibition of viral replication (e.g. 2'-5'-oligo-adenylate synthetase), activate a serine/inhibition ofviral replication (e.g. 2'-5'-oligo-adenylate synthetase), produce variants within an infected host which are selected for survival by selective pressure. The continuous generation of new variants by antigenic drift is leading to an irreversible decline in immune function and, ultimately, to its collapse. Some viruses have evolved very sophisticated ways of avoiding recognition by lymphocytes. For example, adenoviruses may protect infected cells against recognition by cytotoxic T cells by suppressing expression of major histocompatibility class I (MHC-I) proteins, which are essential for recognition of viral peptides by these cells. Other viruses can manipulate cytokine pathways to counteract destruction by the host immune response. Epstein-Barr virus codes for an analog of mammalian IL-10 and stimulates a Th2 response to avoid induction of inflammatory responses. Another strategy is based on the production of materials that interfere with the production of proinflammatory cytokines. For example, vaccinia and cowpox viruses can code for a soluble glycoprotein that is similar in amino acid sequence to the IL-1 receptor and acts as a competitor for secreted IL-1. Vaccinia virus also encodes homologs of both epidermal and endothelial growth factors. Vaccinia and herpes simplex viruses have developed methods of interfering with the alternate and/or classical pathways of complement activation by encoding proteins that function as receptors for complement components including C3b. Herpesviruses such as varicella-zoster and herpes simplex virus encode Fc receptors that can bind immunoglobulins. Poxviruses encode several proteins that have the capacity to interfere with components of the interferon pathway, a well-recognized inhibitor of virus ofreplication. Measles virus, Epstein-Barr virus, and HIV have been shown to modulate levels of surface expression integrins such as LFA-1, LFA-3, and ICAM-1, which are considered as crucial mediators of cell-to-cell interactions. Such a virus-induced change in the expression of integrins may facilitate dissemination of the virus throughout the host or help these viruses avoid immune surveillance. It appears that viruses utilize the components of the immune system for their own purposes in at least four ways:

1. Viruses use cell surface proteins, known to play crucial roles in immunity, as their primary receptors. For example, rhinoviruses utilize as portal of entry ICAM-1, an adhesion molecule that is, through its interaction with LFA-1, important for lymphocyte adherence.

2. Some viruses can escape the normal immune response by using cells of the immune system as hosts. Herpesviruses and retroviruses are the two best-known groups of viruses able to productively infect specific cellular subsets of the immune network. For example, the main target cells for the human herpesvirus 6 are T and B lymphocytes, while the primary cellular reservoirs for HIV remain CD4-expressing T lymphocytes and monocytes/ macrophages. The ability of lymphocytes to divide and yet to remain in a resting state for long periods of time makes them an ideal microenvironment for viruses. Other viruses can create a larger pool of cells serving as putative targets for viral propagation (e.g., mouse mammary tumor viruses). This type of strategy is achieved by the production of virus-encoded superantigens, proteins that stimulate a much larger proportion of T and/or B cells than the nominal antigen.

3. Components of the immune system may be used by viruses to upregulate their replication. A good example of this is the observation that the transcriptional activity of the promoter region of HIV is upregulated by several cytokines (e.g., TNF-α , IL-1, and IL-6).

4. Enveloped viruses can acquire host-encoded proteins that will jeopardize elimination of the virion by the immune response.[17] This is exemplified by the demonstration that acquisition of host-derived MHC-I and ICAM-1 proteins by cytomegalovirus and HIV, respectively, can confer protection against neutralization by antibodies.

In conclusion, viral pathogenesis appears as dynamic and highly complex. The course of viral diseases is determined by a balance between viral and host factors, in which use or misuse of the immune system play a major role. In some diseases such as measles or type B hepatitis, the symptoms of an infection are not directly caused by the pathogen, but are part of the immune response.

Chapter 4

VERTEBRATE VIRUSES: INFECTION

4.I. TYPES OF DISEASE

The general picture of virus diseases resembles that of an iceberg. A vast majority of contacts without infection or of subclinical infections constitutes the invisible part and only a relatively few infections result in clinical disease or death. To some extent, the same pattern exists on the cellular level.

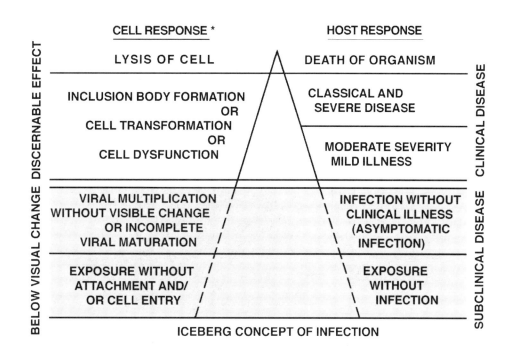

FIGURE 1

"Iceberg" concept of infectious diseases at level of the cell and at level of the host. Within any population, varying patterns of cell response also occur. *Hypothetical. (From Evans, A.S., in *Viral Infections of Humans,* 3rd ed., Evans, A.S., Ed., Plenum Press, New York, 1982, 3. With permission.)

Human infections are respiratory, faecal-oral, genital, or congenital, by saliva, contact, or breach of the skin. Diseases implicating animals, namely vertebrate or invertebrate vectors or reservoirs, are called zoonoses. The infecting virus remains at the body surface, causing a local infection, or enters the body and causes a generalized infection via bloodstream, lymphatics, or nerves. Many viruses have a particular target organ. Viral impact on the host depends largely on the tissue involved. For example, poliovirus infection of the intestine leads to little if any disease, but infection of motor neurons leads to paralysis. Some organs such as the liver have considerable functional reserves which aid in overcoming virus infections. Viral pathogenesis appears as a dynamic process in which viral and host factors balance each other.

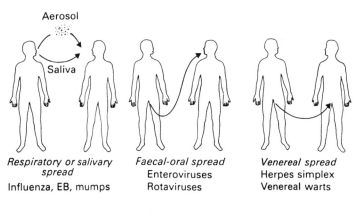

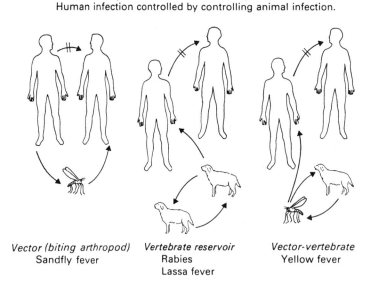

FIGURE 2

Types of transmission of viral infection in man. Respiratory or salivary spread - not readily controllable. Faecal-oral spread - controllable by public health measures. Venereal spread - control socially difficult. Zoonoses - human infection controlled by controlling vector, or by controlling animal infection. (From Mims, C.A. and White, D.O., *Viral Pathogenesis and Immunology,* Blackwell Scientific Publications, Oxford, 1984, 72. With permission.)

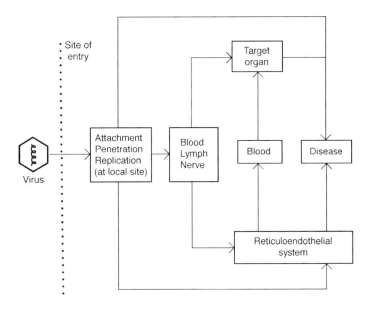

FIGURE 3

Viral spread in the organism. (From Stringfellow, D.A., in *Virology*, Stringfellow, D.A., Ed.-in-chief, Upjohn, Kalamazoo, 1983, 45. Courtesy Pharmacia & Upjohn.)

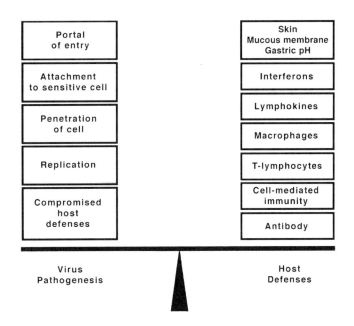

FIGURE 4

Balance between viral pathogenesis and host defenses. (From Stringfellow, D.A., in *Virology*, Stringfellow, D.A., Ed.-in-chief, Upjohn, Kalamazoo, 1983, 45. Courtesy Pharmacia & Upjohn.)

Acute infection is usually followed by virus clearance, although viruses or virus parts (measles virus nucleocapsids, virus genomes in herpes virus infections) may persist and cause late complications or relapses. Slow progressive infections are mainly represented by prion-caused subacute spongiform encephalopathies.

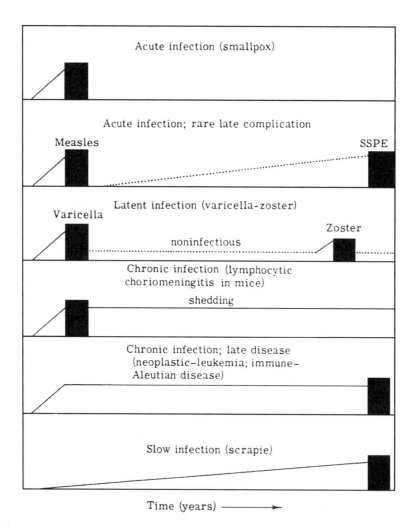

FIGURE 5

Acute and various kinds of persistent infections. Solid line, demonstrable infectious virus; dotted line, virus not readily demonstrable; box, disease episode. SSPE, subacute sclerosing panencephalitis. (From Fenner, F., McAuslan, B.R., Mims, C.A., Sambrook, J., and White, D.O., *The Biology of Animal Viruses,* 2nd ed., Academic Press, New York, 1974, 453. With permission.)

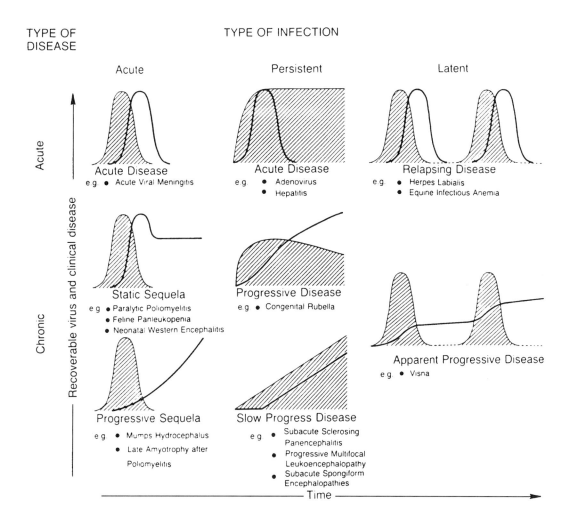

FIGURE 6

Patterns of infection and disease. Acute, persistent, and latent infections are shown by shaded areas. The course of clinical disease is shown by a solid line. (From Johnson, R.T., Narayan, O., and Clements, J., in *Persistent Viruses,* Stevens, J.G, Todaro, G.J., and Fox, C.F., Eds., Academic Press, New York, 1978, 551. With permission.)

On the cellular level, virus infections are productive, latent, abortive, or restrictive with partial expression of viral genes. Productive infections may be lytic, resulting in cell death with release of novel progeny virions. In steady-state infections, viruses are released by budding and the cell survives for some time. This is seen in many enveloped viruses. Integrated infections result from persistence of viral genomes either as integrated DNA or as episomes.

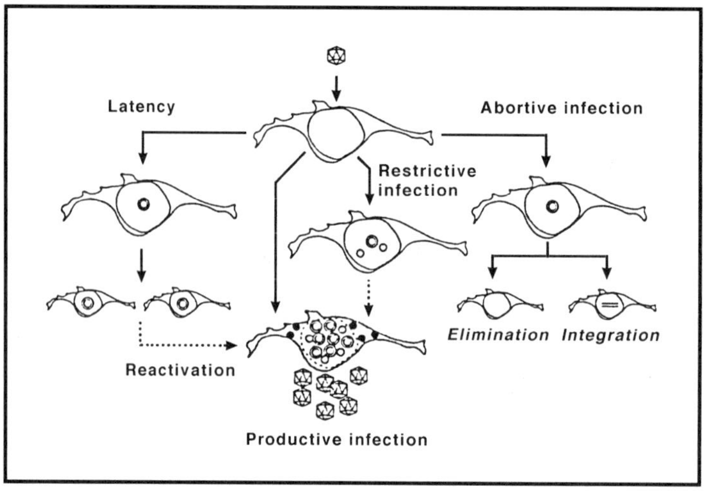

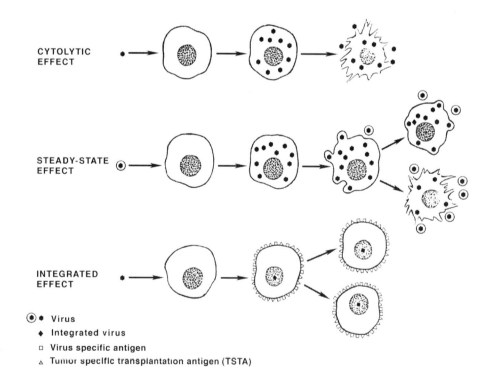

FIGURE 7

Types of infection and virus-cell interactions. (Adapted from Maréchal, V., Dehée, A., and Nicolas, J.-C., *Virologie,* 1 (special issue), 11, 1997. With permission of John Libbey Eurotext.)

FIGURE 8

Three types of virus-host cell interactions. The drawing is limited to eukaryote viruses and does not account for (a) filamentous phages of the fd type which induce steady-state infections and (b) integrated viruses which occasionally cause cell lysis. (Author's note) (From Bellanti, J.A., *Immunology,* W.B. Saunders, Philadelphia, 1971, 271. With permission.)

Many acute virus diseases share a pattern exemplified by the following diagram illustrating a localized infection. Virus and interferon titers peak early in disease and antibodies rise much later when clinical symptoms already abate. This pattern of acute infections dictates the time to collect specimens for virus isolation (early in the acute phase) and serum (in the acute and convalescent phases).

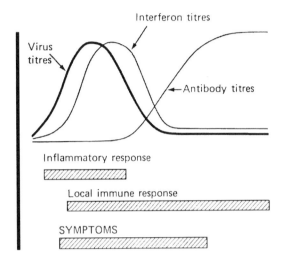

FIGURE 9

Events occurring in a typical localized infection. In these simple infections it will be noted that viral application is restricted to the portal of entry of the virus and it will be seen that clinical symptoms do not appear until the final stages of the growth cycle. (From Robinson, T.W.E. and Heath, R.B., *Virus Diseases and the Skin,* Churchill Livingstone, Edinburgh, 1983, 23. With permission.)

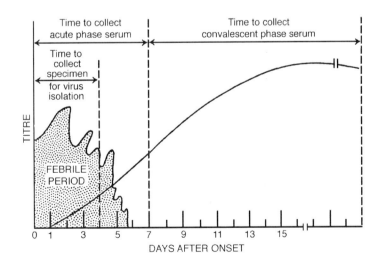

FIGURE 10

Nature and course of the disease influence the timing of medical interventions such as collecting specimens. (Author's legend) (From Anonymous, *Viral and Rickettsial Diseases, Physician's Handbook,* 4th ed. Ontario Department of Health, Toronto, 29. © 1972 Queen's Printer for Ontario. Reproduced with permission.)

4.II. VIRUS SPREAD

The mucosal membranes of the respiratory, gastrointestinal, and genitourinary tracts are the most important portals of entry. Here naked viable cells are exposed to the environment with little protective covering. Mucous membranes are also sites of virus shedding to the exterior. Virus penetration through the skin requires abrasions or wounds and is achieved by bites of vertebrates or arthropods (mosquitoes, ticks, sandflies) or human intervention (e.g., blood transfusion or needle inoculation). Many viruses remain confined to the site of entry. This is the case in many infections of the upper respiratory tract, in papillomas and warts, and in rotavirus gastrointestinal infections. Viruses causing generalized infections reach their target organs via the bloodstream, lymph vessels, or nerves. Hematogenous spread is frequently biphasic, involving several stages of viral replication and viremia. Viruses are free in the plasma or vehiculated by blood cells (e.g., erythrocytes, platelets, lymphocytes, monocytes/macrophages).

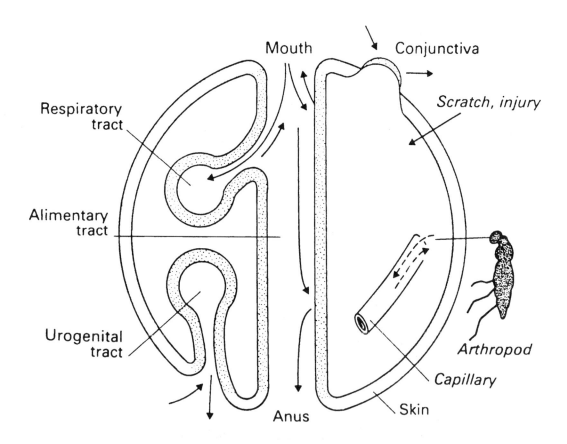

FIGURE 11

Body surfaces as sites of virus infection and shedding. (From Mims, C.A. and White, D.O., *Viral Pathogenesis and Immunology*, Blackwell Scientific Publications, Oxford, 1984, 40. With permission.)

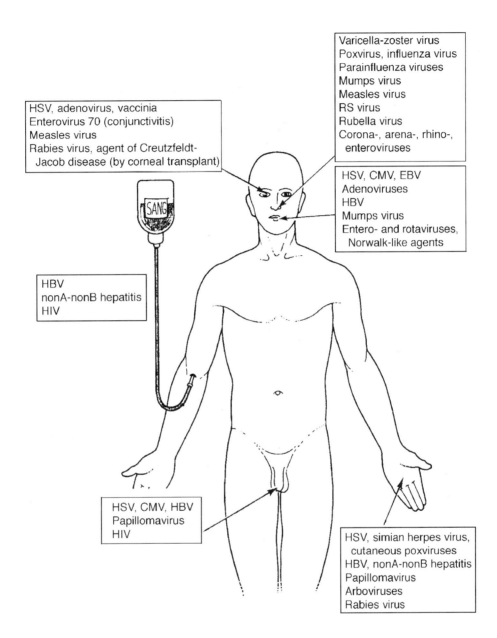

FIGURE 12

Entry sites of the most important viruses. (From Huraux, J.M., Nicolas, J.C., and Agut, H., *Virologie,* Flammarion Médecine Sciences, Paris, 1985, 21. With permission.)

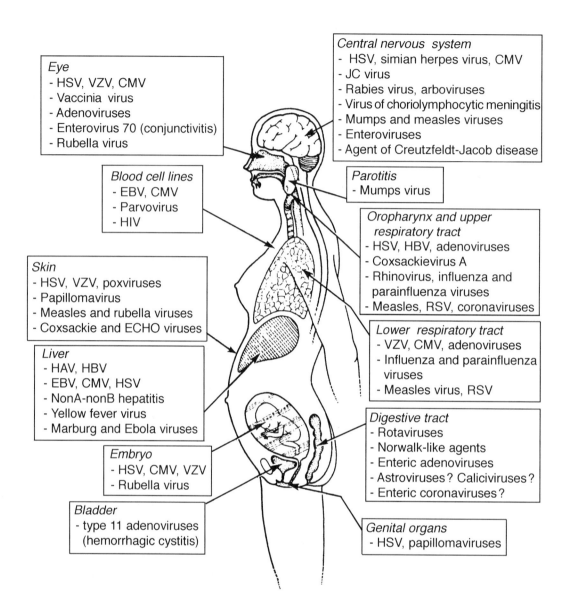

FIGURE 13

Target organs of the most important viruses. A target is any organ whose infection causes the characteristic signs of the viral disease. (From Huraux, J.M., Nicolas, J.C., and Agut, H., *Virologie,* Flammarion Médecine Sciences, Paris, 1985, 23. With permission.)

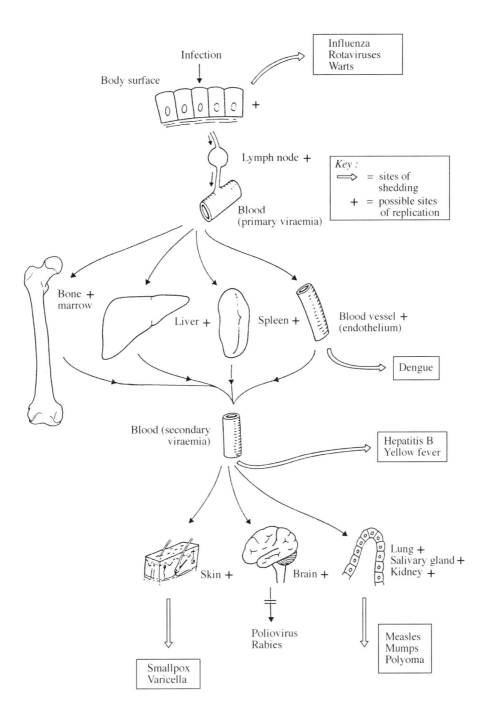

FIGURE 14

The spread of infection through the body. (Modified from Mims, C.A. and White, D.O., *Viral Pathogenesis and Immunology,* Blackwell Scientific Publications, Oxford, 1984, 52. With permission.)

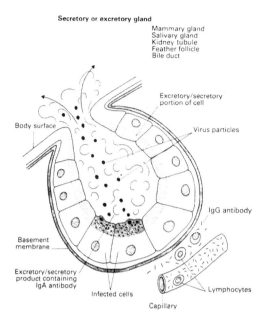

FIGURE 15

Viral infection of cell surfaces facing the exterior. (From Mims, C.A. and White, D.O., *Viral Pathogensis and Immunology,* Blackwell Scientific Publications, Oxford, 1984, 246. With permission.)

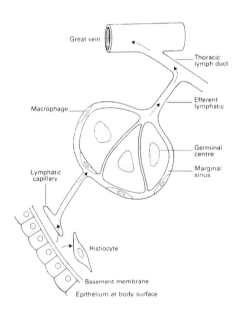

FIGURE 16

Subepithelial invasion and lymphatic spread of viruses. (From Mims, C.A. and White, D.O., *Viral Pathogenesis and Immunology,* Blackwell Scientific Publications, Oxford, 1984, 50. With permission.)

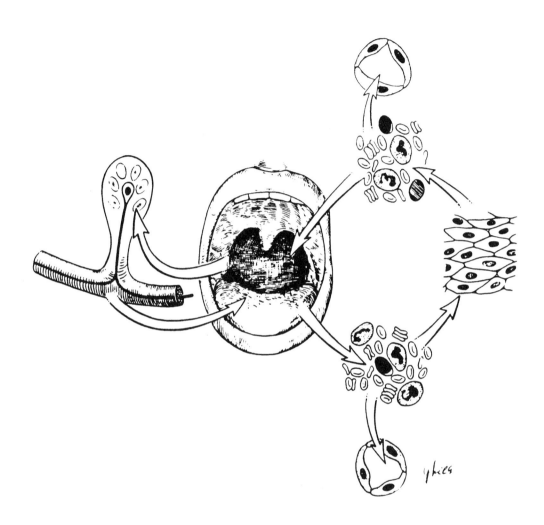

FIGURE 17

Pathogenesis of oropharyngeal infections showing hematogenous *(right)* and neural *(left)* spread of virus from and to the oropharynx. In hematogenous spread, there may be initial growth within specific cells of the oropharynx or other sites. Viruses in the blood are removed by the reticuloendothelial system. To maintain a viremia, other sites of virus growth are usually established with amplification of the viremia, which may seed 1 virus to susceptible cells within the oropharynx. Neural spread may occur from sensory nerve endings in the lip, oral mucosa, or tonsillar tissue, along nerves with cell bodies in the sensory ganglia. If latency is established within cells in the ganglia, virus may be activated later with spread down the sensory nerves to susceptible cells in the oropharynx. (From Johnson, R.T., in *Viral Infections in Oral Medicine,* Hooks, J. and Jordan, G., Eds., Elsevier/North-Holland, New York, 1982, 3. With permission of R.T. Johnson.)

Most viral infections of the CNS are acquired from the blood via capillaries of the brain or the choroid plexus. The neural route leads mainly to infection of spinal and cranial nerve ganglia.

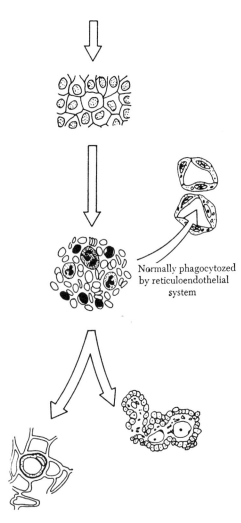

1. Entry into host
    Enteric
    Respiratory
    Inoculation (animal or
        arthropod bite)

2. Growth in extraneural tissues
    A. Primary sites
        Gastrointestinal or respiratory
            tracts
        Subcutaneous tissue or vascular
            endothelium
        Lymph nodes

    B. Secondary sites
        Liver, spleen, bone marrow,
        muscle, connective tissue,
        brown fat.

3. Maintenance of viremia
    Sufficient input
    Adsorption to red cells
    Growth in white cells
    Inactive reticuloendothelial
        system

4. Crossing from blood to brain
    A. Choroid plexus to CSF
        Passage through choroid plexus
        Growth in choroid plexus

    B. Small vessels to brain
        Transport by white cells
        Growth through vessels
        Diffusion through vessels
        Across areas of permeability
        Across normal capillaries or
        venules

Normally phagocytozed
by reticuloendothelial
system

FIGURE 18

Steps in the hematogenous spread of viruses to the central nervous system (courtesy of R.T. Johnson). (From Fenner, F., *The Biology of Animal Viruses,* Vol. 2, *The Pathogenesis and Ecology of Viral Infections,* Academic Press, New York, 1969, 517. With permission of R.T. Johnson.)

Neural spread is centripetal or centrifugal to and from the central nervous system (CNS). It is observed notably in rabies and herpesvirus infections. Viruses may move within the axon, perineural lymphatics, endoneural spaces, or Schwann cells. Viruses can cross neuromuscular junctions and move within synapses.

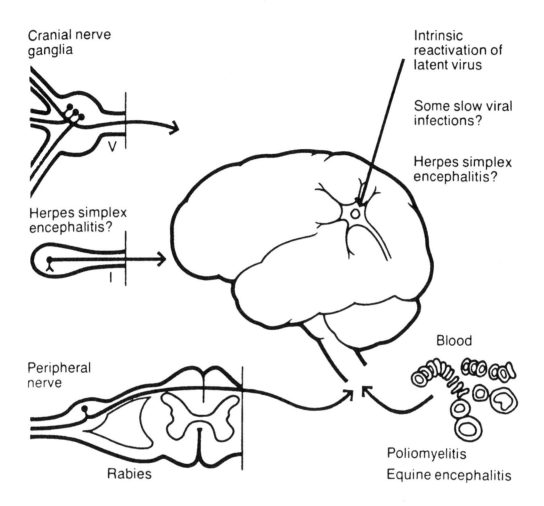

FIGURE 19

Pathways of viral entry into the CNS. Primary viral replication may occur in skin, muscle, or gastrointestinal or respiratory tract. (From Bale, J.F. and Kern, E.R., in *Virology,* Stringfellow, D.A., Ed., Upjohn, Kalamazoo, 1983, 77. Courtesy Pharmacia & Upjohn.)

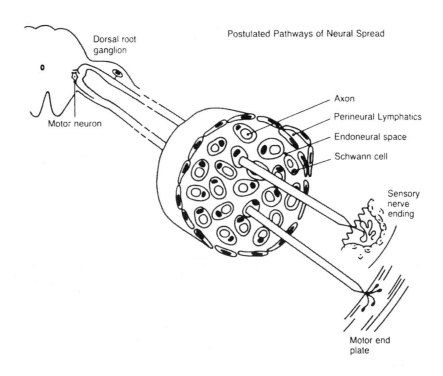

FIGURE 20

Neural pathways for CNS infection. Virus can be taken up at sensory and motor endings and moved within axons, endoneural space, or perineural lymphatics or by Schwann-cell infection. If movement is via axons, viruses taken up at sensory endings will be delivered selectively to dorsal root ganglia and those at motor endings to motor neurons. (From Johnson, R.T., *Viral Infections of the Nervous System*, Raven Press, New York, 1982, 44. © Lippincott Williams & Wilkins. With permission.)

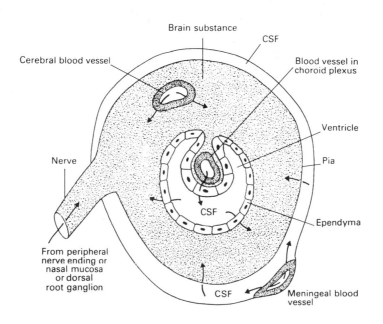

FIGURE 21

Routes of viral invasion of the central nervous system. CSF, cerebro-spinal fluid. (From Mims, C.A. and White, D.O., *Viral Pathogenesis and Immunology,* Blackwell Scientific Publications, Oxford, 1984, 61. With permission.)

Olfactory rods are the only nerve cells whose naked axons are in direct contact with the environment. Although aerosol infections, notably with poliovirus, have shown that viruses can invade the CNS from the olfactory mucosa, the olfactory route does not appear as an important pathway to CNS infections in humans.

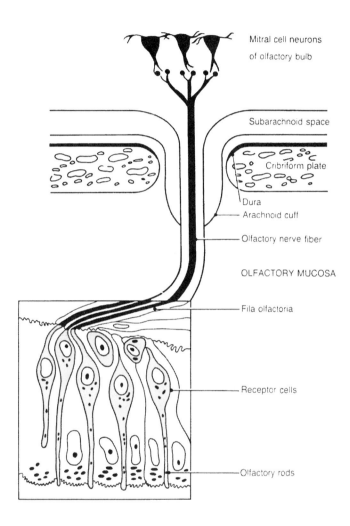

FIGURE 22

Olfactory pathways for CNS infection. The dura is affixed to the inner table of the skull and is penetrated by the arachnoid, which forms a cuff surrounding the olfactory fibers in the nasal submucosal tissue. Receptor cells *(below)*, which synapse directly with mitral cell processes in the olfactory bulb of the CNS, have rods that extend beyond the epithelium of the nasal mucosa. (Modified from Johnson, R.T. and Griffin, D.E., in *Handbook of Clinical Neurology,* Vol. 34, Part II., Vinken, P.J., Bruyn, G.W., and Klawans, J.H.L., Eds., North-Holland Publishing Company, Amsterdam, 1978, 15; from Johnson, R.T., *Viral Infections of the Nervous System,* Raven Press, New York, 1982, 47. © Lippincott Williams & Wilkins. With permission.)

Viruses may infect the fetus via the ovum, amniotic fluid, or the placenta. Infection is by complete virions or, transovarially, integrated viral genomes. Transovarial infection is seen in many retroviruses (e.g., murine leukemia, mouse mammary tumors, and avian leukosis). In mammals, viral infection results in a wide spectrum of diseases ranging from inapparent infections to fetal death with abortion or resorption of the fetus, fetal disease with or without malformations, or neonatal disease. The outcome depends often on the time of infection; for example, human rubella in the first stages of pregnancy leads to malformations. Congenital or genetic infection of chickens by avian leukosis virus generally leads to inapparent infections with or without viremia.

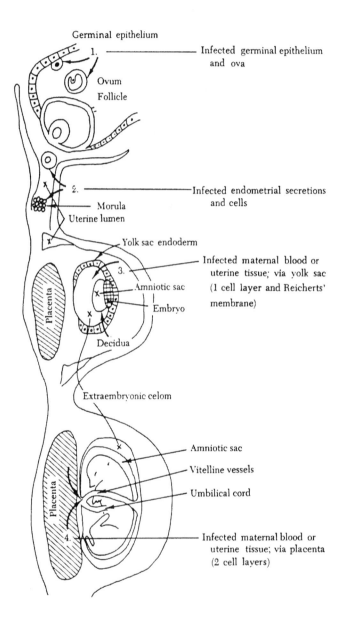

FIGURE 23

The reproductive system of the mouse embryo and possible routes of infection of the embryo by viruses. In other mammals, the situation is similar, but (a) the yolk sac is very variable in different mammals, and (b) the placental junction (number of cell layers and their permeability) varies in different species of mammals and at different stages of pregnancy. (From Fenner, F., McAuslan, B.R., Mims, C.A., Sambrook, J., and White, D.O., *The Biology of Animal Viruses,* 2nd ed., Academic Press, New York, 1974, 384. With permission.)

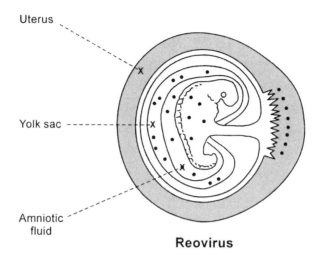

Uterus

Yolk sac

Amniotic
fluid

Reovirus

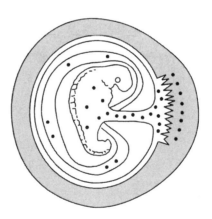

Parvovirus

FIGURE 24

Fetal viral infection via the placenta: in reoviruses by way of amniotic fluid (A), and in parvoviruses mainly by way of umbilical cord (B). Dots represent viruses. (By Ackermann, developed from Kilham, G. and Margolis, G., *Progr. Med. Virol.*, 20, 113, 1975.)

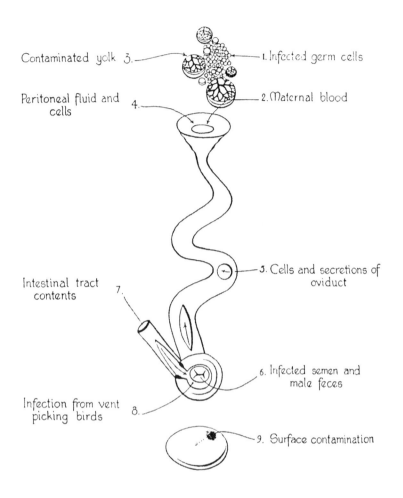

Contaminated yolk 3. —— 1. Infected germ cells

Peritoneal fluid and cells 4. —— 2. Maternal blood

Intestinal tract contents 7. —— 5. Cells and secretions of oviduct

—— 6. Infected semen and male feces

Infection from vent picking birds 8. —— 9. Surface contamination

FIGURE 25

The hen's reproductive system and ways disease agents may gain entrance into eggs.

1. *Infected germ cell.* The female germ cells may carry disease agents into eggs.

2. *Maternal blood.* Some birds have a slight hemorrhage following ovulation. Thus, maternal blood, which could carry disease agents, is often included into the egg, forming the so-called blood spot.

3. *Contaminated yolk.* The yolk material accumulates over a period of several weeks before ovulation. Disease agents could be deposited in the yolk and thus carried into the egg.

4. *Peritoneal fluid and cells.* Peritoneal infections can be carried via the infundibulum by means of peritoneal fluid and cells.

5. *Cells and secretions of oviduct.* The egg spends about 25 hours in the oviduct of the hen. During this time the chalazae, the albumen, shell membranes, shell and cuticle are added to the egg. An infection from the oviduct cells could be easily included into the egg.

6. *Infected semen and male feces.* The seminal fluid and spermatozoa could carry disease agents. In addition, feces from male birds occasionally enter the female during copulation.

7. *Intestinal tract contents.* In many hens a physiological prolapse of the oviduct occurs at the time the egg is laid. This allows the everted wall of the oviduct to come in contact with the cloacal lining and contamination can thus be taken into the oviduct as it is retracted.

8. *Infection from vent-picking birds.* Contamination of the bills can be transferred to the oviduct of other birds that are the victims of cannibalism. The vent-picking habit is quite common in laying hens, especially in groups which contain hens that are slow to retract their oviduct after laying.

9. *Surface contamination* and penetration of the shell after the egg is laid. (From Cottrall, G.E., *Ann. N.Y. Acad. Sci.,* 55, 221, 1952. With permission.)

On the cellular level, viruses spread (1) from cell to cell in the extracellular milieu, (2) from cell to cell by intercellular bridges or cell fusion, or (3) vertically as genomes which are, whether integrated or as plasmids, passed to daughter cells. Viral spread may involve several cell types; for example, hepatotropic viruses in the liver are captured by and propagated in Kupffer cells, a part of the reticuloendothelial system, and later passed to hepatocytes.

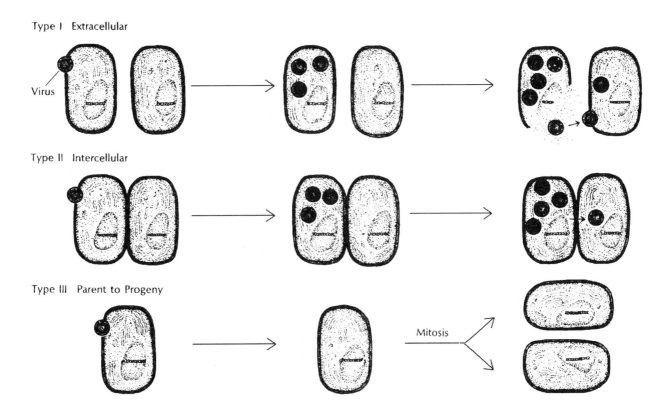

FIGURE 26

Three routes of viral spread on the cell level are recognized. In type I the infected cell is lysed and new viruses spread extracellularly to near and distant uninfected cells. Viruses that spread by the type II route induce cell fusion and then spread intercellularly. In type III spread, the virus is incorporated into the cell genome and passes to progeny during mitosis. (Reproduced with permission. From Notkins, A.I., Viral infections: mechanisms of immunologic defense and injury. *Hospital Practice,* 9 (9), 65, © 1974 The McGraw-Hill Companies, Inc. Illustration by Nancy Lou Riccio.)

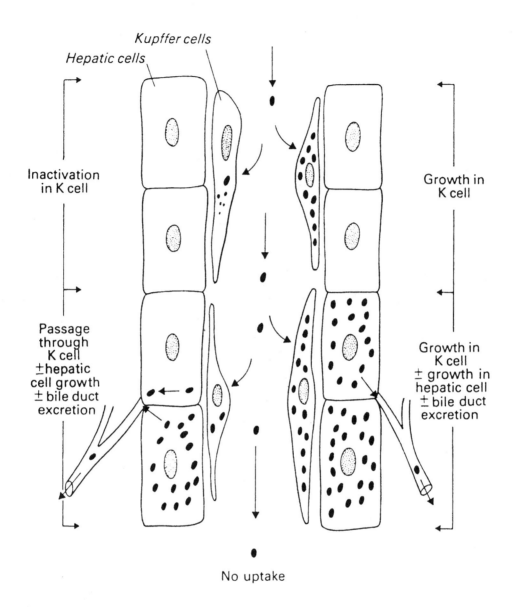

FIGURE 27

Types of viral behaviour in liver. Endothelial cells have been omitted; in most infections their role is unknown. (From Mims, C.A. and White, D.O., *Viral Pathogenesis and Immunology*, Blackwell Scientific Publications, Oxford, 1984, 55. With permission.)

4.III. CYTOPATHOLOGY

Many vertebrate viruses produce cytopathic effects in cell cultures and *in vivo*. Cells may become rounded and shrunken, detach themselves from the substrate, lyse, or form syncytia or intranuclear or intracytoplasmic inclusion bodies. These morphological changes are valuable for preliminary virus identification in culture.

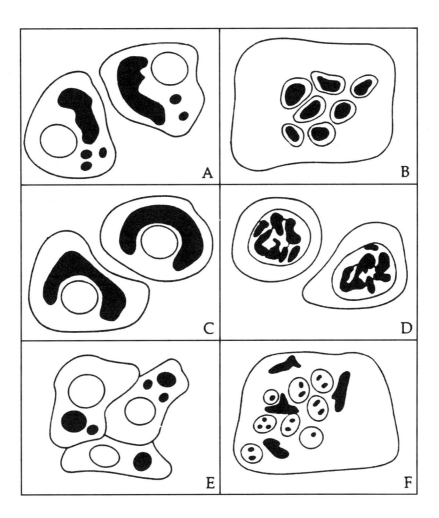

FIGURE 28

Inclusion bodies in virus-infected cells. (A) Vaccinia virus - intracytoplasmic acidophilic inclusion (Guarnieri body, B body). (B) Herpes simplex virus - intranuclear acidophilic inclusion (Cowdry type A); cell fusion produces syncytium. (C) Reovirus - perinuclear intracytoplasmic acidophilic inclusion. (D) Adenovirus - intranuclear basophilic inclusion. (E) Rabies virus - intracytoplasmic acidophilic inclusions (Negri bodies). (F) Measles virus - intranuclear and intracytoplasmic acidophilic inclusions; cell fusion produces syncytium. (From Fenner, F., McAuslan, B.R., Mims, C.A., Sambrook, J., and White, D.O., *The Biology of Animal Viruses*, 2nd ed., Academic Press, New York, 1974, 342. With permission.)

Most families of dsDNA viruses and the retroviruses include oncogenic viruses, able to induce tumors *in vivo*. In retroviruses, this requires transcription of viral RNA into dsDNA and integration of the latter into the host genome. Cultured cells infected by oncogenic viruses either produce novel viruses or are transformed into an oncogenic phenotype. Transformation is characterized by absence of virus production, persistence of the viral genome (or part of it), and alteration of cellular growth (e.g., loss of contact inhibition, formation of microtumors). Simultaneously, viral proteins (T or "tumor" antigens) may be expressed at the cell surface.

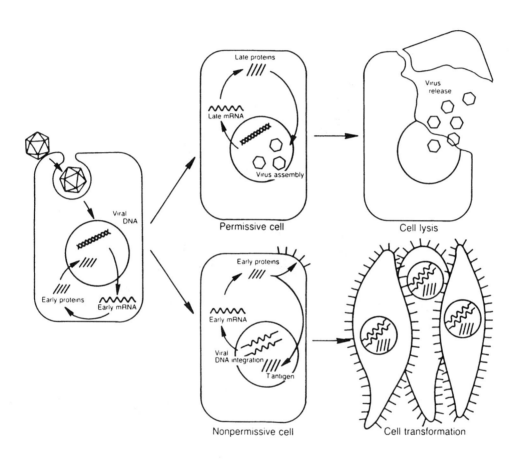

FIGURE 29

Cell transformation by papovaviruses and adenoviruses. Icosahedral viruses are engulfed into the cell. Viral DNA is delivered to the nucleus, and early messenger RNA directs protein synthesis in the cytoplasm, including the tumor (T) protein which accumulates in the nucleus. In the permissive cell *(upper panels),* late messenger RNA is transcribed, and late proteins are returned to the nucleus, where the virus is assembled. Cell ruptures to release progeny virions. In the nonpermissive cell *(lower panels),* all or part of the viral DNA is integrated, with continued production of early messenger RNA, accumulation of T antigen in the nucleus, and alternations of the cytoplasmic membrane causing cell transformation without production of progeny virions. (From Johnson, R.T., *Viral Infections of the Nervous System,* Raven Press, New York, 1982, 298. © Lippincott Williams & Wilkins. With permission.)

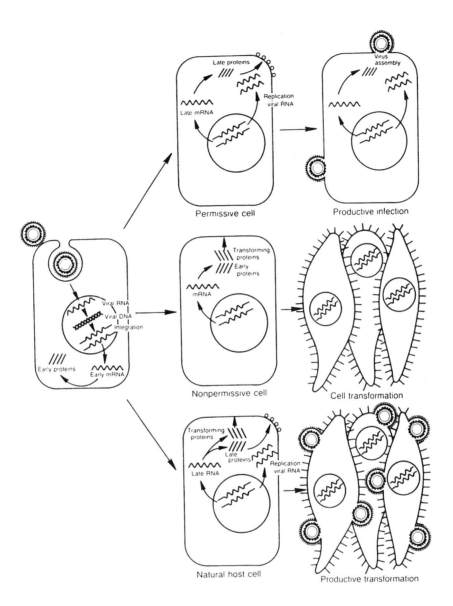

FIGURE 30

Cell transformation by retroviruses. Retrovirus penetrates the cell, and the viral RNA is released into the cell, where proviral RNA is replicated and integrated into the cell DNA *(left)*. Subsequently, messsenger RNA is transcribed from integrated proviral DNA. In the permissive cell *(upper panels),* viral structural proteins are translated, and viral RNA is replicated, with assemblage of virus at the cytoplasmic membrane and budding of virions at the cytoplasmic membrane and budding of virions. In the nonpermissive cell *(center panels),* viral genes that determine transformation direct production of transforming proteins. These proteins in the cytoplasm and cytoplasmic membrane alter cell growth, but no progeny virus is produced. In cells of the natural host *(lower panels),* permissive infection and transformation may occur simultaneously. (From Johnson, R.T., *Viral Infections of the Nervous System,* Raven Press, New York, 1982, 299. © Lippincott Williams & Wilkins. With permission.)

On the molecular level, virus infection alters intracellular signalling pathways and stimulates or down-regulates growth factors, cytokines, chemokines, and apoptosis. Another consequence is increased production of "stress proteins". These constitutive proteins facilitate virus replication and assembly. When expressed on the surface of virus-infected cells, they also mediate elimination of these cells by the immune system.

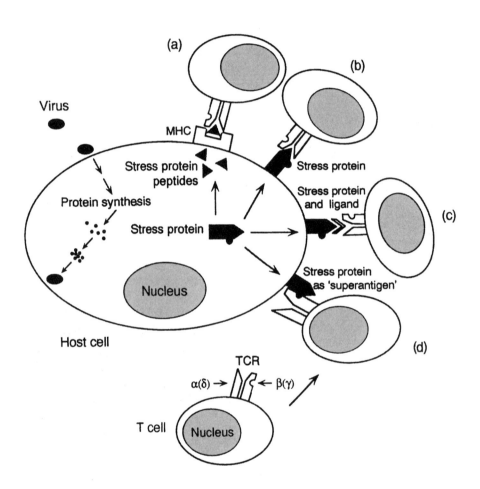

FIGURE 31

Virus infection of a host cell leads to a stress response, resulting in increased synthesis of stress proteins (hsps). Stress proteins have several putative roles to play within virus-infected cells. They may facilitate the assembly and replication of the virus. In addition, the infected cells may also display stress proteins or their determinants on the cell surface, resulting in the activation of the immune system. T cells may recognize (a) a complex hsp-peptide-MHC (major histocompatibility complex); (b) hsp itself, using both chains of the T cell receptor (TCR) dimeric molecule; (c) a viral or nonviral peptide-hsp complex; or even (d) hsp as "superantigen" through a particular single chain of the the TCR dimer. Published data can be interpreted in terms of possibilities (b), (c), and (d). (Reprinted from *Trends Microbiol.*, 2, 89-91, 1994. Jindal, S. and Malkowsky, M., Stress responses to viral infection. © 1994, with permission of Elsevier Science.)

Apoptosis, or programmed cell death, is characterized by cell shrinkage, DNA degradation, and appearance of cell fragments or "apoptotic bodies" which are phagocytosed. This is different from necrosis, a pathological form of cell death accompanied by inflammation. Apoptosis is a normal phenomenon required for tissue homeostasis. Viruses may induce apoptosis, leading to elimination of infected cells, or inhibit it in a variety of ways.

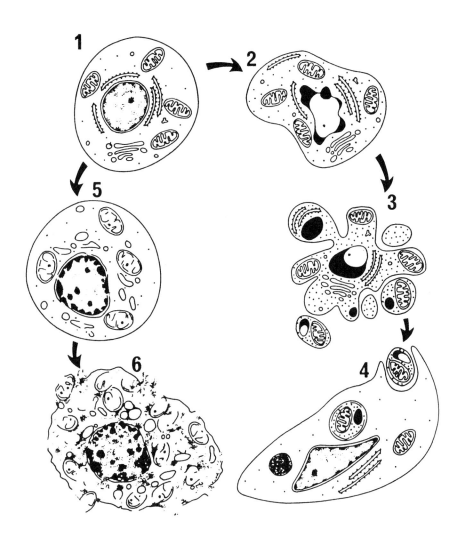

FIGURE 32

Diagram illustrating the sequential ultrastructural changes in apoptosis *(right)* and necrosis *(left)*. A normal cell is shown at (1). The onset of apoptosis (2) is heralded by compaction and segregation of chromatin into sharply delineated masses that lie against the nuclear envelope, condensation of the cytoplasm, and mild convolution of the nuclear and cellular outlines. Rapid progression of the process over the next few minutes (3) is associated with nuclear fragmentation and marked convolution of the cellular surface with the development of pedunculated protuberances. The latter then separate to produce membrane-bound apoptotic bodies, which are phagocytosed and digested by adjacent cells (4). Signs of early necrosis in irreversibly injured cell (5) include clumping of chromatin into ill-defined masses, gross swelling of organelles, and the appearance of flocculent densities in the matrix of mitochondria. At a later stage (6) membranes break down and the cell disintegrates. (From Kerr, J.F.R. and Harmon, B.V., in *Apoptosis: The Molecular Basis of Cell Death. Current Communications in Cell and Molecular Biology,* Vol. 3, Tomei, L.D. and Cope, F.O., Eds., Cold Spring Harbor Laboratory Press, Cold Spring Harbor, 1991, 5. With permission.)

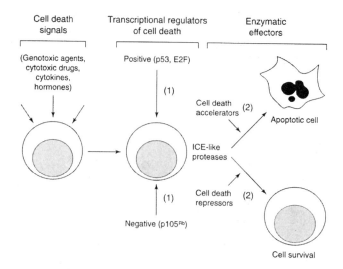

FIGURE 33

Apoptosis may be inhibited by viral infection. Apoptosis is controlled by a balance of factors acting, upstream, as transcriptional regulators of apoptosis pathways (1) and, downstream, as effectors of programmed cell death (2). Effectors include members of the Bcl-2 gene family such as *bax* and *bak* (accelerators) and *bcl-2* or *bcl-X$_l$* (repressors). (Reprinted from *Trends Microbiol.,* 4, 312-316, 1996. Gillet, G. and Brun, G., Viral inhibition of apoptosis. © 1996, with permission of Elsevier Science.)

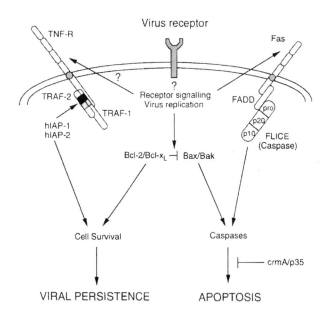

FIGURE 34

Relationships between cellular and viral proteins that modulate apoptosis in virus infection. Alphaviruses can induce the host cell to undergo apoptosis. The death pathway is dependent on viral and cellular factors. The cellular *bax* and *bak* genes, known to accelerate cell death, also accelerate virus-induced apoptosis. Suppression of apoptosis by the product of the *bcl-2* gene can facilitate a persistent virus infection. (From Griffin, D.E. and Hardwick, J.M., *Annu. Rev. Microbiol.,* 51, 565, 1997. With permission from the *Annual Review of Microbiology,* © 1997 by Annual Reviews http://www.AnnualReviews.org.)

Chapter 5

VERTEBRATE VIRUSES: HOST DEFENSE

5.I. ORGANISMIC DEFENSE

The first and most formidable barrier against viruses is the skin. The outer layer of the epidermis consists of dead keratinized cells which do not replicate viruses, while its inner layer has neither blood vessels, lymphatics, or nerves. Mucosal membranes are protected by IgA-containing, virus-trapping mucus and contain polarized epithelial cells which reject some viruses into the bronchial or gut lumen. In the respiratory tract, special ciliated cells move mucus and foreign particles including viruses outward. In addition, the relatively low temperature of the upper respiratory tract inhibits many viruses. In the gastrointestinal tract, most swallowed viruses are inactivated by the acidic pH of the stomach, proteases, and bile salts which disrupt viral envelopes. This explains why, except for coronaviruses, enteric viruses are not enveloped. The fetus is protected by fetal membranes and the placenta; however, viruses can be carried across, leak through, or grow across these barriers.

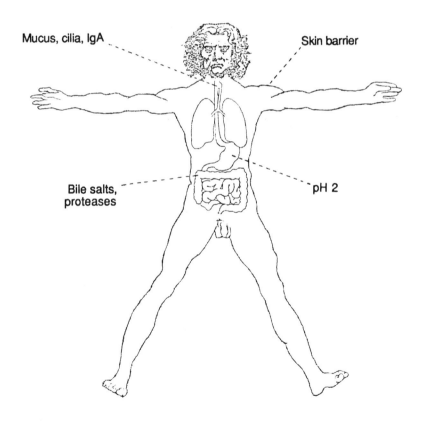

FIGURE 35
The first defenses against viral infection are located at external and internal body surfaces. (By Ackermann, derived from a study of body proportions by Leonardo da Vinci, 1485-1490, Venice, Galleria dell'Accademia.)

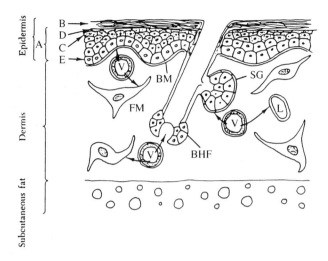

FIGURE 36

Structures of the skin which are of importance in the pathogenesis of viral lesions. The epidermis consists of a layer several cells thick of living cell, the Stratum malpighii (A), covered by a layer of dead keratinized cells, the Stratum corneum (B). The upper cells of the Stratum malpighii contain increasing amounts of keratin granules; sometimes two narrow layers, the Stratum granulosum (C) and the Stratum lucidum (D), are distinguished. The basal layer of dividing epidermal cells is the Stratum germinativum (E). The dermis contains blood vessels (V), lymphatic vessels (L), and fibroblasts and macrophages (FM). The ground substance contains collagenous, reticular, and elastic fibers. The hair follicle (BHF) and sebaceous glands (SG) are appendages of the epidermis. (From Fenner, F., McAuslan, B.R., Mims, C.A., Sambrook, J., and White, D.O., *The Biology of Animal Viruses,* 2nd ed., Academic Press, New York, 1974, 347. With permission.)

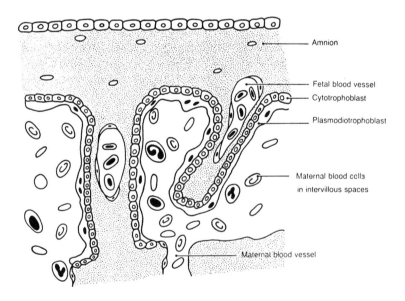

FIGURE 37

Relationship of maternal blood and fetal blood within the human placenta. Maternal blood flows from maternal vessels (below) into open intervillous spaces. Fetal blood (with nucleated red cells) flows through vascular villi separated from the maternal circulation by the trophoblast cells and syncytia of fetal origin. (From Johnson, R.T., *Viral Infections of the Nervous System*, Raven Press, New York, 1982, 205. © Lippincott Williams & Wilkins. With permission.)

The lymphoid system constitutes the inner defense. It consists of a central component (bone marrow, thymus, spleen) and a peripheral part (spleen, lymph nodes and vessels, unencapsulated scattered tissues such as the tonsils and Peyer's patches in the intestine). The thymus and its equivalent in birds, the bursa of Fabricius, direct immunogenesis in the young and are responsible for cell-mediated immunity. Lymph nodes remove foreign particles and recirculate lymphocytes. The spleen, among other functions, produces lymphocytes and plasma cells. Lymphocytes are constantly recirculated.

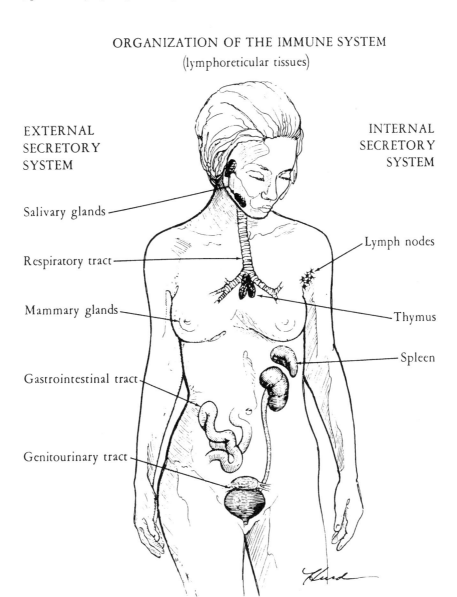

ORGANIZATION OF THE IMMUNE SYSTEM
(lymphoreticular tissues)

EXTERNAL SECRETORY SYSTEM

INTERNAL SECRETORY SYSTEM

Salivary glands

Respiratory tract

Mammary glands

Gastrointestinal tract

Genitourinary tract

Lymph nodes

Thymus

Spleen

FIGURE 38

Location of lymphoreticular tissues. The system consists of cells distributed strategically over the body as well as lining lymphatic and vascular channels. Its cells are housed within thymus, lymph nodes, spleen (internal secretory system), and those body tracts exposed to the external environment, the respiratory, gastrointestinal, and genitourinary tracts (external secretory systems). (From Bellanti, J.A., *Immunology*, W.B. Saunders, Philadelphia, 1971, 17. With permission.)

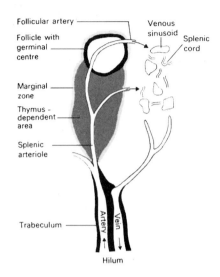

FIGURE 39

The human spleen. Lymphoid cells form a sheath around the arterioles (white pulp). The remainder (red pulp) consists of splenic cords and venous sinusoids filled with erythrocytes. (From Roitt, I.M., *Essential Immunology,* 4th ed., Blackwell Scientific Publications, Oxford, 1980, 83. With permission.)

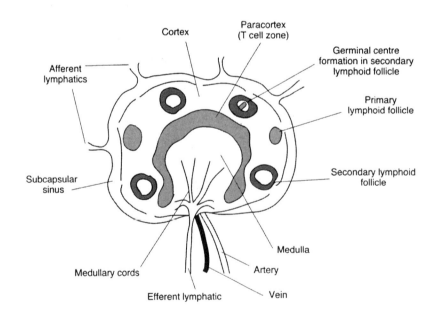

FIGURE 40

Transverse section of a lymph node. (Reprinted from *Trends Microbiol.,* 3, 434-440, 1995. Blacklaws, B., Bird, P., and McConnell, I., Early events in infection of lymphoid tissue by a lentivirus, maedi-visna. © 1995, with permission of Elsevier Science.)

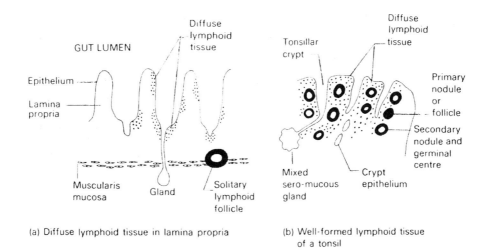

FIGURE 41

Unencapsulated lymphoid tissue. (From Roitt, I., *Essential Immunology,* 4th ed., Blackwell Scientific Publications, Oxford, 1980, 84. With permission.)

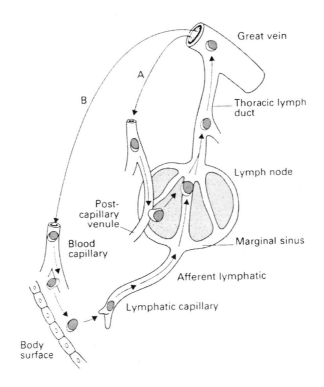

FIGURE 42

Lymphocyte recirculation. Recirculating lymphocytes in man are mostly T lymphocytes; approximately 90% of recirculation is by route A and 10% is by route B. (From Mims, C.A, and White, D.O., in *Viral Pathogenesis and Immunology,* Blackwell Scientific Publications, Oxford, 1984, 90. With permission.)

5.II. CELLULAR DEFENSE

NONSPECIFIC AND SPECIFIC IMMUNITY

Defense on the cellular level is mediated by the immune system. Immune responses are nonspecific (innate, natural) or specific (induced) for particular pathogens or antigens. Both are mediated by humoral (soluble) factors or cellular elements. Nonspecific immunity is provided by phagocytic macrophages in blood and tissue, "natural killer" or NK cells, the complement and properdin systems, and cytokines including interferon. Macrophages eliminate free viruses and other particulate material by phagocytosis. NK cells are sentinels that kill virus-infected cells. Complement can inactivate certain viruses in the absence of antibodies. Nonspecific immunity is aided by fever and local acidity or hypoxia. Specific immunity is provided by antibodies and sensitized cytotoxic lymphocytes.

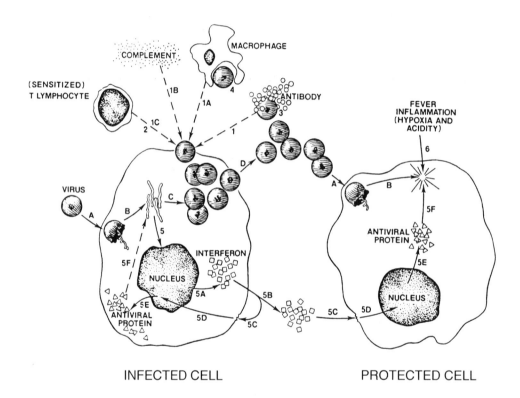

FIGURE 43

Host defenses against virus infection. The extracellular defenses include antibody (1, 3) and the phagocytic macrophage (4) which may combine with and inactivate virus. Defenses against intracellularly replicating virus include: cytolysis of virus-infected cells bearing virus surface antigens by synthetized T lymphocytes (2) acting directly or through their mediators; antibody plus macrophages (1A); antibody plus complement (1B); antibody plus T lymphocytes (1C); and antibody plus K cells (not shown). Additional inhibition of intracellular viral replication (D and C) can be caused by the nonspecific defense mechanisms of interferon (5-5F), local acidity, local hypoxia (6), and interfering virus (not shown). (From Baron, S. and Dianzini, F., in *The Interferon System: A Current Review to 1978*, Baron, S. and Dianzini, F., Eds., *Texas Rep. Biol. Med.*, 35, 1, 1977. With permission.)

The immune system of animals developed over a long time. The first components to appear, already present in invertebrates, seem to be phagocytosis and inflammation. These primitive responses are supplemented in vertebrates by specific immunity and amplification systems (kinins, complement). A specific immunologic response appears in the form of disseminated cells in the hagfish. Subsequently, specialized lymphoid organs appear, such as the thymus or the bursa of Fabricius in birds. Antibody production appears late and diversifies along with vertebrate evolution. The most primitive vertebrates produce an immunoglobulin of the M class; other classes appear later during evolution. Lymphoid cells and hematopoietic cells derive from common undifferentiated stem cells in the bone marrow.

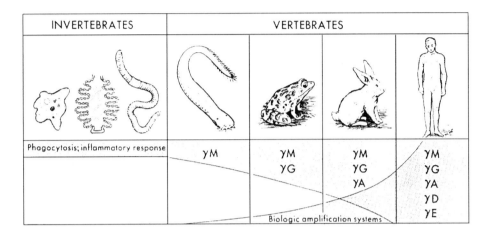

A

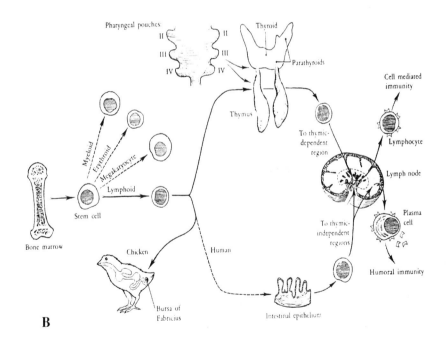

B

FIGURE 44

A. Phylogeny of the immune response. B. Ontogeny of immune response with differentiation of progenitor cells into hematopoietic and immunocompetent cells. (From Bellanti, J., *Immunology*, W.B. Saunders, Philadelphia, 1971, 64 and 67. With permission.)

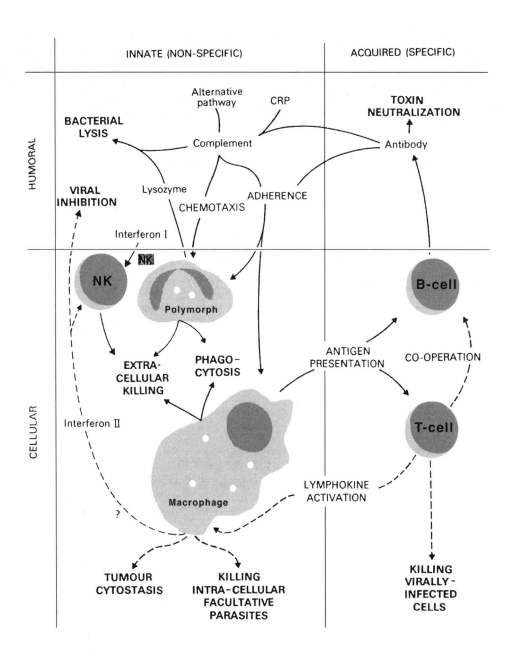

FIGURE 45

Interactions between natural and specific immunity mechanisms. Reactions influenced by T cells are indicated by a broken line. (Developed from Playfair, J.H.L., *Brit. Med. Bull.,* 30, 24, 1974; from Roitt, I.M., *Essential Immunology,* 4th ed., Blackwell Scientific Publications, Oxford, 1980, 215. With permission by J.H.L. Playfair and Blackwell Science.)

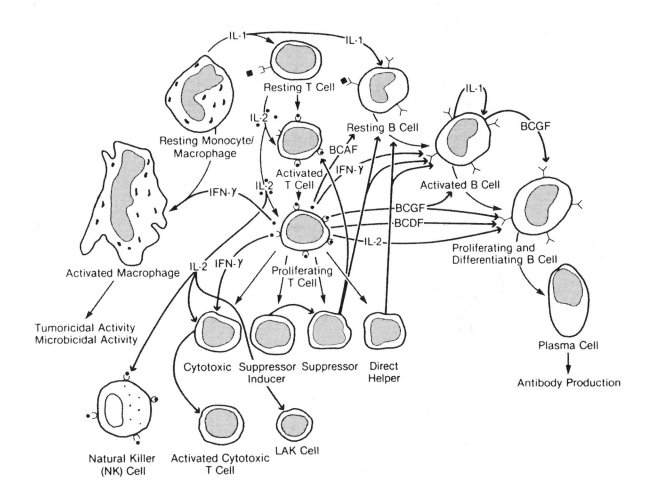

FIGURE 46

The human immunoregulatory network. IL, interleukin; IFN, interferon; LAK, lymphokine-activated killer cell; BCGF, B-cell differentiation factor; BCAF, B-cell activating factor. (From Fauci, A.S., Rosenberg, S.A., Sherwin, S.A., Dinarello, C.A., Longo, D.L., and Lane, H., *Ann. Int. Med.,* 106, 421, 1987. With permission.)

INTERFERON

"Interferon" is a family of cytokine proteins produced by vertebrate cells in response to virus infection and other stimuli. Type I interferon (IFNα, IFNβ) is made by virus-infected cells of any type. Type II (IFN γ) is produced by T lymphocytes and NK cells. Both have multiple functions. Interferons are host species-specific rather than virus-specific. Upon release by virus-infected cells, interferons bind to specific receptors of other cells and trigger there a synthesis of antiviral proteins that inhibit viral replication by several mechanisms. Interferons are induced immediately after viral infection, but disappear rapidly.

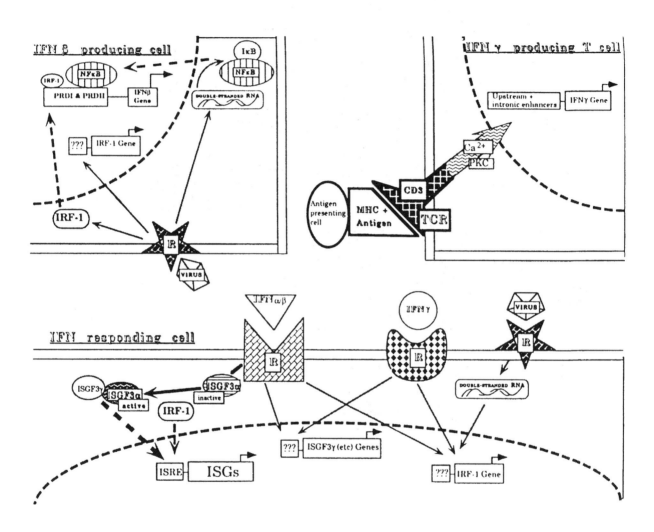

FIGURE 47

Overview of the interferon system. Virus interaction with the IFN-β-producing cell *(upper left)* via a receptor (R) generates active IRF-1 and NFκB-like factors, by an undefined mechanism. Together with others, these factors mediate transcriptional activation of the IFN- β gene. After immune recognition of virus-derived antigens, T cells *(upper right)* produce IFN-γ. These IFNs exert effects on target cells *(lower)* through interaction with specific receptors, reulting in the production of activated transcription factors such as ISGF-3 and IRF-1. Additionally, viruses interacting with IFN-exposed cells also induce activation of transcription factors. The latter accumulate in the nucleus, inducing transcription of the ISG (IFN-stimulated genes), which encode antiviral activities. TCR, T lymphocyte antigen receptor; CD3, T lymphocyte CD3 complex. (From Sen, G.C. and Ransohoff, R.M., *Adv. Virus Res.,* 42, 57, 1993. With permission.)

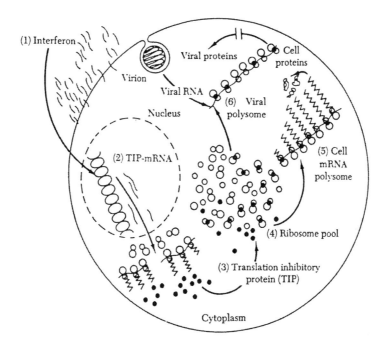

FIGURE 48

An early model of interferon action. A few molecules of interferon attached to the cytoplasmic membrane (1) act to derepress a host cell cistron, initiate transcription of the messenger RNA, TIP-mRNA (2), which encodes for the translation-inhibitory protein (TIP) that in turn is synthetized and accumulates in the cytoplasm (3), where it binds to ribosomes and contributes TIP+-units to the ribosome pool (4). Polysomes composed of TIP+-ribosomes and cellular mRNA are translated normally (5), whereas polysomes formed from TIP+-ribosomes and viral RNA are not translated (6), producing a state of interferon-mediated interference. (From Marcus, P.I. and Salb, J.M., *Virology*, 30, 502, 1966. With permission.)

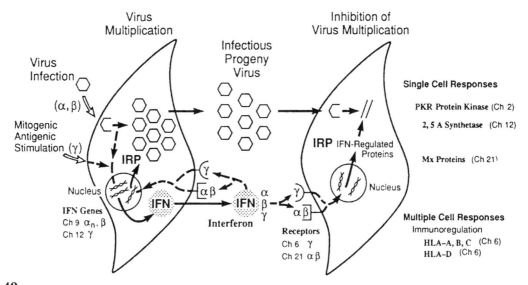

FIGURE 49

The interferon system (IFN) based on the study of human cells. At left, a virus-infected (stimulated) cell produces IFN in response to virus infection (α- and β-IFNs) or mitogens and antigens (γ-IFN) The IFN-treated cell at right responds to the presence of IFN by producing new proteins that blocks virus replication. Some IFN-regulated proteins (IRP), responsible for inhibition of virus replication within single cells (PKR protein kinase, 2',5'-oligoadenylate synthetase, Mx proteins) or within whole animals via multiple cell responses (histocompatibility antigens, HLA) are listed on the right. The chromosome (Ch) assignments are for human cells. (From Samuel, C.E., in *Encyclopedia of Virology*, Vol. 2, Lederberg, J., Ed.-in-chief, Academic Press, San Diego, 1992, 533. With permission.)

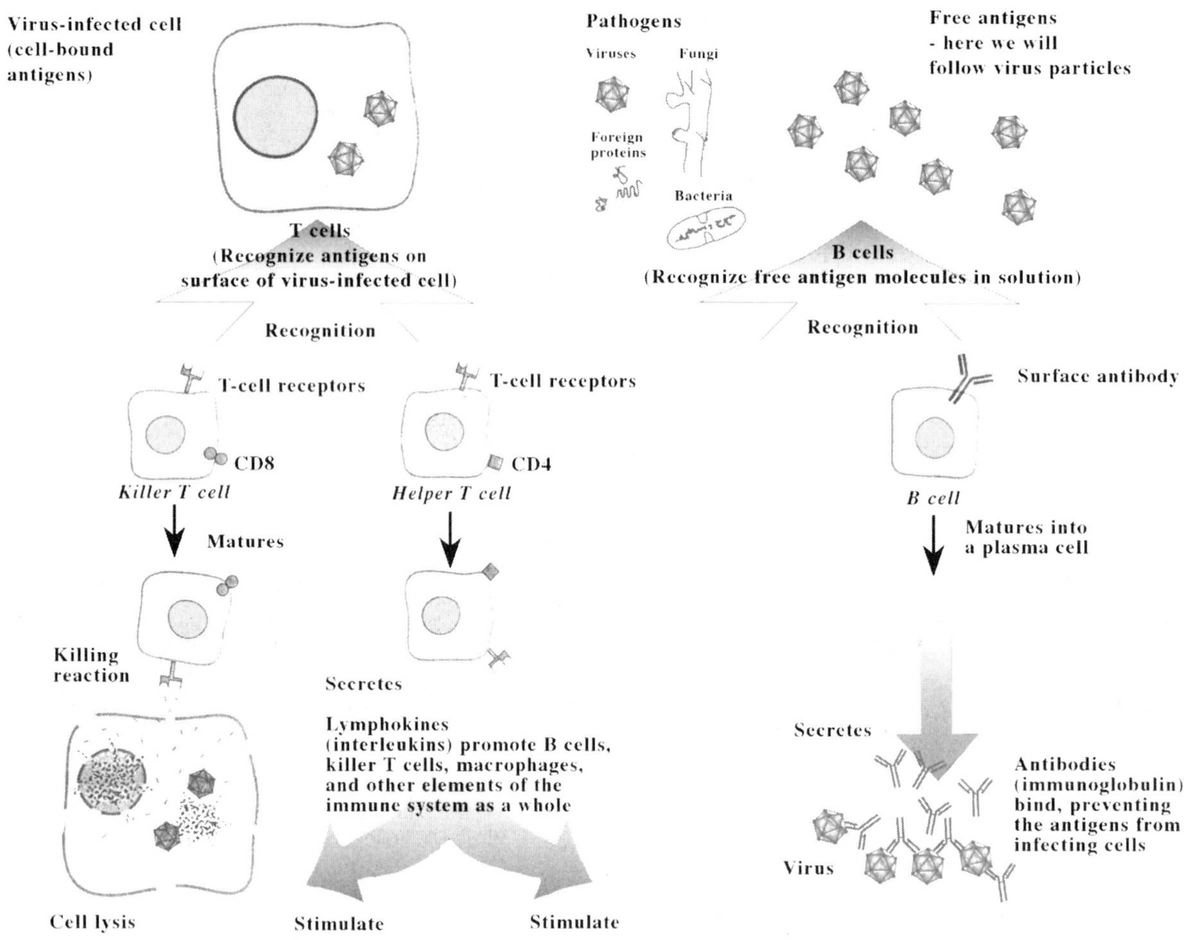

FIGURE 50

The two arms of the immune response, cellular and humoral (serum) immunity:

I. *Cellular immunity* is a consequence of the activity of T cells (killer and helper T cells). Both have T-cell receptors on their surfaces and recognize cell-bound molecules on plasma membranes of infected cells. They neither respond to free antigens in body fluids nor secrete antibodies. (1) The *Killer T cell* has CD4 receptors and take infected cells (not free antigens) as its target. It binds to them and kills them, inactivating the pathogen, here a virus. By boring holes and inducing cell lysis, one killer T cell can prevent virus replication in many host cells and slow virus spread in the organism. (2) Helper T cells have CD4 receptors. Response is the secretion of lymphokines, or interleukins. These feed back into the immune system overall by stimulating both B-cell and T-cell killer response: hence the helper T cell is a key factor for optimal function of both arms of the immune response.

II. *Humoral immunity* (B cell secretions) is a consequence of antibodies, serum proteins that are collectively called immunoglobin and constitute about one-fifth of the proteins in the blood. B cells, recognizing free antigens in the body fluids, mature into plasma cells and secrete antibody molecules that bind to antigens amd inactivate them. Viruses, with many identical binding sites for antibodies (because of their structural symmetry), can form large aggregates when they react with immunoglobin. (From Levine, A.J., *Viruses*, 55. © 1992 by Scientific American Library, New York. Used with permission by W.H. Freeman and Company.)

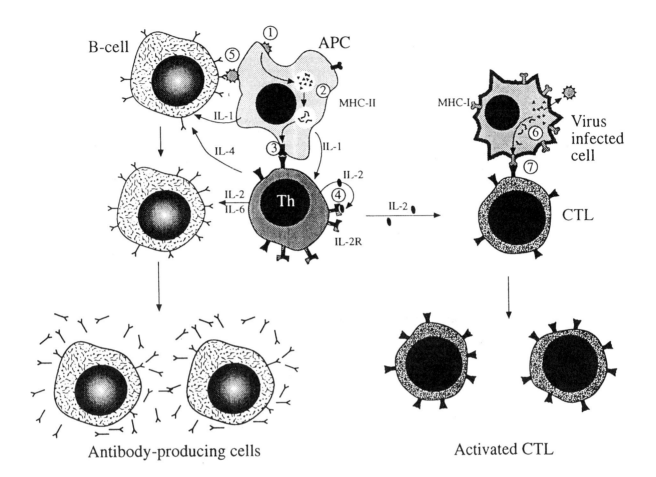

FIGURE 51

Simplified scheme of immune response stimulation. Antigen-presenting cells (APC) take up and process antigen (1) into short peptides which are processed (2) and presented with the MHC class II molecule (3) to antigen-specific T-helper cells (Th). This recognition, plus the stimulation by APC-produced IL-1 causes the Th cell to produce both IL-2 (4) and the IL-2 receptor (IL-2R). Antigen adsorbed to the surface of the APC may also be recognized by specific B cells (5) which are then stimulated, after receiving help in the form of IL-2, IL-4, and IL-6, to progress to antibody-producing cells. Cytotoxic T cells (CTL) are stimulated by specific viral antigen endogenously processed (6) by a virus-infected cell in the context of MHC class I molecules (7). Also after receiving help in the form of IL-2, the CTL proliferates and differentiates to form mature CTL. (From Norley, S. and Kurth, R., in *The Retroviridae,* Vol. 3., Levy, J.A., Ed., Plenum Press, New York, 1992, 363. With permission.)

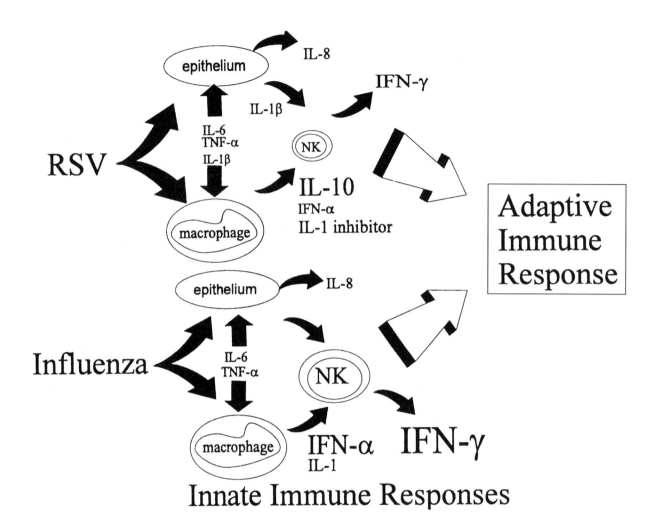

FIGURE 52

RSV (respiratory syncytial virus) and influenza trigger cytokine release by cells resident in the respiratory tract, and the resulting cytokine milieu inhibits subsequent immune events. Influenza induces a strong IFN-α response that stimulates natural killer (NK) cells and a cytokine environment that promotes Th1 differentiation. RSV is a relatively weak inducer of IFN-α, but induces an IL-1 inhibitor, and is a potent inducer of IL-10. This may theoretically result in a relatively weak NK cell response. The balance of these secreted cytokines may be an important determinant of the adaptive immune response. Cytokine release from epithelial cells and macrophages may also influence susceptibility of neighboring cells to viral infection, and may thereby influence disease expression. (From Maletic Neuzil, K. and Graham, B.S., *Sem. Virol.,* 7, 255, 1996. With permission.)

In primo-infections, viruses meet macrophages and NK cells. They induce interferon, chemotactic factors that attract more macrophages, and later production of cytotoxic lymphocytes (CTLs) and antibodies. CTLs recognize virus-infected cells and destroy them early in the replication cycle before new progeny viruses appear. The CTL response peaks after 7-10 days and antibodies rise over a period of 2-4 weeks, but may persist for life. Antibodies neutralize free virus and promote cell lysis by complement. The diagram below illustrates reinfection or an advanced primo-infection.

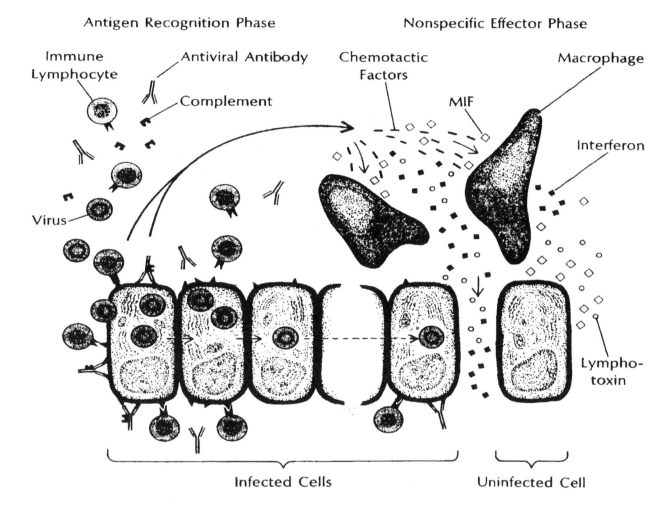

Antigen Recognition Phase   **Nonspecific Effector Phase**

Immune Lymphocyte   Antiviral Antibody   Chemotactic Factors   Macrophage

Complement   MIF   Interferon

Virus   Lympho-toxin

Infected Cells   Uninfected Cell

FIGURE 53

Host immune response to viral infection consists of two phases. In antigen recognition phase, antibody, complement, and immune lymphocytes react with virus-infected cells *(left)*. While this immune response will eventuate in lysis of infected cells, the lysis may not occur until after intracellular transmission of the viral infection has taken place. However, the antigen-recognition phase also leads to the generation of immunologic mediators. In the second, nonspecific phase *(right)*, mediators attract macrophages to the site of infection where they exert toxic effects on both infected and uninfected cells. This toxicity, plus interferon from specifically stimulated lymphocytes, breaks contact between adjacent cells and inhibits viral replication, thus finally halting viral spread. (Reproduced with permission from Notkins, A.I., Viral infections: mechanisms of immunologic defense and injury. *Hospital Practice,* 9 (9), 65. © 1974 The McGraw-Hill Companies, Inc. Illustration by Nancy Lou Riccio.)

HUMORAL IMMUNITY

Humoral immunity is mediated by immunoglobulins (IgG, IgM, IgA) produced by B lymphocytes and their derivatives, plasma cells. The principal virus-antibody reactions are virus neutralization with or without complement, complement-facilitated lysis of viruses or infected cells, and opsonization (coating with antibody to facilitate phagocytosis). Neutralization is mediated by IgG, IgM, and IgA antibodies and affects free viruses. Neutralizing antibodies block the first stages of virus infection. Circulating virus-antibody complexes may localize in the kidney or elsewhere and produce immune diseases such as glomerulonephritis in B hepatitis.

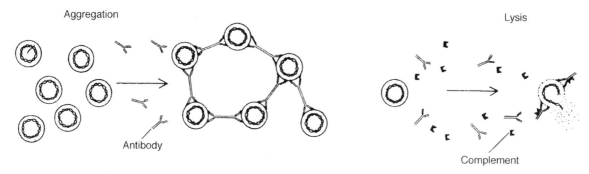

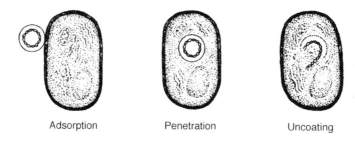

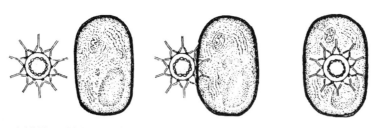

FIGURE 54

Virus neutralization can take place extracellularly or at the cell level. Neutralization occurs extracellularly when antibody aggregates the virus, reducing the number of infectious particles, or when antibody and complement acting together lyse the virus. At the cell level, in the absence of antibody, the virus attaches to the cell surface, penetrates, and uncoats in preparation for replication. However, antibody can block the virus at any of these three steps. (Reproduced with permission. Adapted from Notkins, A.I., Viral infections: mechanisms of immunologic defense and injury. *Hospital Practice,* 9 (9), 65. © 1974 The McGraw-Hill Companies, Inc. Illustration by Nancy Lou Riccio.)

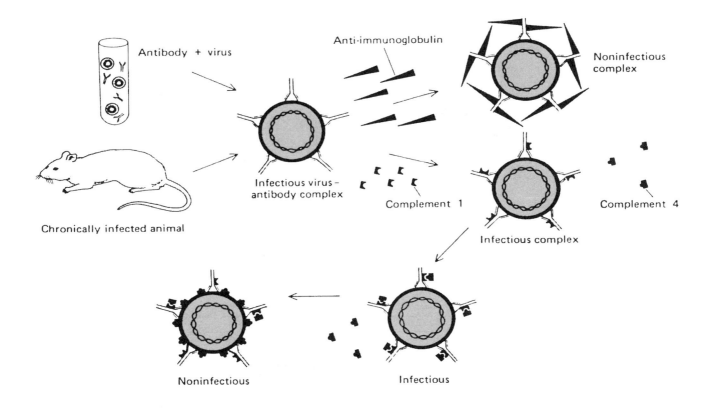

FIGURE 55

Interaction of antibody with virus does not always result in neutralization, and the existence of infectious virus-antibody complexes both *in vivo* and *in vitro* was demonstrated by incubating a virus with anti-immunoglobulin. This attaches only to virus that has antiviral antibody on it, and the result is a more efffective covering for the surface and neutralization. Complement neutralizes virus-antibody complexes in a similar manner. The early components of complement effect neutralization by piling up on the virion. Only C1 and C4 are necessary, but they must interact in that sequence. (Reproduced with permission. Adapted from Notkins, A.I., Viral infections: mechanisms of immunologic defense and injury. *Hospital Practice,* 9 (9), 65. © 1974 The McGraw-Hill Companies, Inc. Illustration by Albert Miller.)

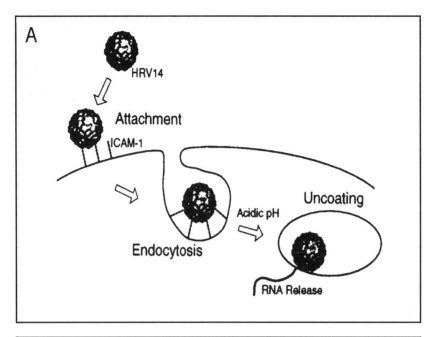

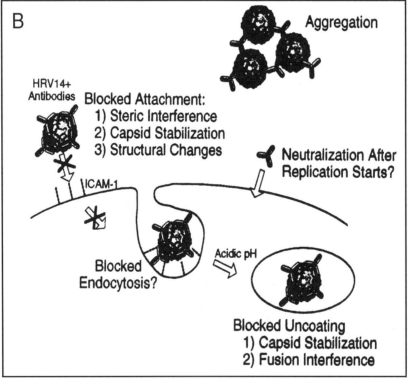

FIGURE 56

Human rhinovirus 14 infection (A) and possible mechanisms of antibody-mediated neutralization (B). (From Smith, T.J., Mosser, A.G., and Baker, T.S., *Sem. Virol.,* 6, 233, 1995. With permission.)

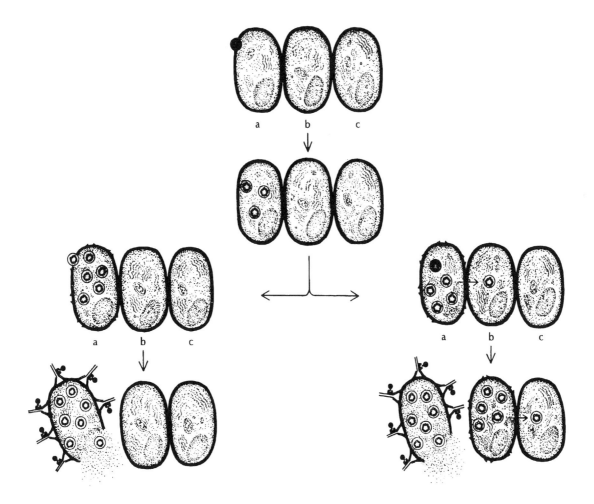

FIGURE 57

Virus-infected cells can be destroyed by antiviral antibody and complement, but whether this halts type II spread depends in part on the speed of transmission. If, for example, the infected cell is destroyed before the virus is transmitted to cell B, the infection would be stopped *(left)*. If, on the other hand, the virus transmission to cell B occurs before cell A can be destroyed immunologically, the infection would progress. (Reproduced with permission. From Notkins, A.I., Viral infections: mechanisms of immunologic defense and injury. *Hospital Practice,* 9 (9), 65. © 1974 The McGraw-Hill Companies, Inc. Illustration by Nancy Lou Riccio.)

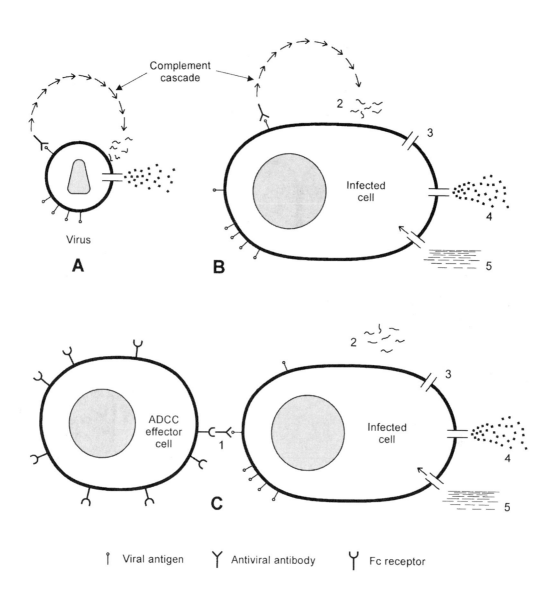

FIGURE 58

Antibody-dependent lysis of an enveloped virus (here a retrovirus) and virus-infected bacteria. (A) In virolysis, antibodies binding to the virus may activate the complement cascade and cause disruption of the virus by the tubular poly-C9 membrane attack complex. In some cases the viral proteins may activate complement directly. (B) Complement-mediated lysis of a cell. (1) Antibodies bind to viral glycoproteins present on the surface of infected cells and activate the complement cascade, leading to the polymerization (2) and insertion of tubular poly-C9 complexes (3). The cell is lysed by osmotic shock involving loss of cellular proteins (4) and entry of surrounding fluid (5). (C) Lysis of infected cells by antibody-dependent cell-mediated cytotoxicity (ADCC). (1) Killer cells with Fc receptors recognize the Fc portion of antibodies against viral glycoproteins present on infected cells. (2) This stimulates the secretion and deposition of perforin molecules which, similar to the C9 component of complement, (3) form a lytic, tubular polyperforin attack complex. The infected cell is lysed as above by loss of cellular proteins (4) and influx of surrounding fluid (5). (By Ackermann, developed from reference 46.)

Secretory IgA antibodies can traverse epithelial membranes and collect in the mucus covering mucosal surfaces. Their function is to stop virus infection at the earliest stage.

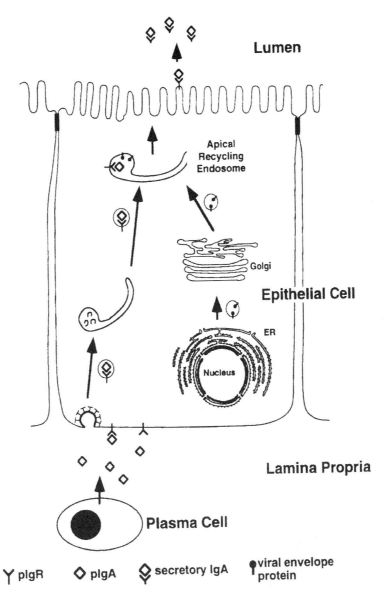

FIGURE 59

Functioning of IgA antibodies in the mucosal epithelium. Plasma cells in the *Lamina propria* of a mucous membrane secrete polymeric IgA (pIgA) which binds to its receptor, pIgR, on the basolateral surface of an epithelial cell. The complex is endocytosed and transcytosed in vesicles to the apical surface, where the pIgR is cleaved, leaving secretory IgA in the lumen. Antigens in the *Lamina propria* can be bound by IgA antibody and excreted in immune complexes through the epithelium by the same route as free IgA. Virus infecting an epithelial cell can theoretically be neutralized intracellularly if transcytosing IgA antibody is able to bind to a viral component and interfere with virus synthesis or assembly. In the hypothetical scheme shown, viral envelope protein that was synthetized in the rough endoplasmic reticulum (ER) and processed in the Golgi is transported in a post-Golgi vesicle to an apical recycling endosome that contains transcytosing antiviral IgA, which then disrupts viral assembly. A final level of action for free secretory IgA antibody in the lumen is to bind antigens, thus providing an immune exclusion barrier that inhibits access to the epithelium. (From Lamm, M.E., *Annu. Rev. Microbiol.*, 51, 311, 1997. With permission from the *Annual Review of Microbiology*, © 1997, by Annual Reviews http://www.AnnualReviews.org.)

Reinfection or infection after vaccination lead to a rapid, massive IgG response called "anamnestic reaction." The reason is that part of B lymphocytes are converted into memory B cells, perhaps through the persistence of viral antigen.

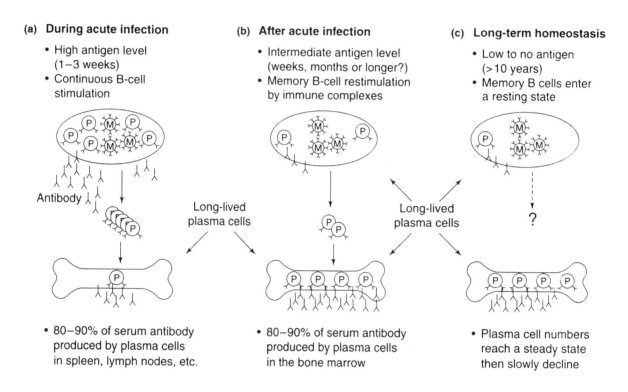

(a) During acute infection
- High antigen level (1–3 weeks)
- Continuous B-cell stimulation

Antibody

Long-lived plasma cells

- 80–90% of serum antibody produced by plasma cells in spleen, lymph nodes, etc.

(b) After acute infection
- Intermediate antigen level (weeks, months or longer?)
- Memory B-cell restimulation by immune complexes

Long-lived plasma cells

- 80–90% of serum antibody produced by plasma cells in the bone marrow

(c) Long-term homeostasis
- Low to no antigen (>10 years)
- Memory B cells enter a resting state

?

- Plasma cell numbers reach a steady state then slowly decline

FIGURE 60

Model of long-term antibody production. (a) During the initial viral infection, B cells become stimulated by high levels of antigen to become virus-specific antibody-secreting cells. Most serum antibody is produced in the spleen (lymph nodes, etc.) and few plasma cells accumulate in bone marrow. (b) After resolving the viral infection, most antigen is bound in antigen-antibody complexes on follicular dendritic cells and may intermittently stimulate memory B cells to differentiate into plasma cells. At this stage, most serum antibody is produced by long-living plasma cells in the bone marrow, with only a few plasma cells remaining in the spleen. It is possible that terminal differentiation of plasma cells occurs after migration to the bone marrow. (c) Months after an acute infection, antigen levels may drop below an immunological threshold of activation and the immune response would then return to homeostasis. Memory B cells enter a resting state and no longer differentiate into plasma cells, unless stimulated by antigen. The model proposes that plasma cells in the bone marrow, and a subpopulation of plasma cells in the spleen and lymph nodes, have a long life span, and that serum antibody levels may be sustained for several in the absence of antigenic restimulation. M, memory B cells; P, plasma cells. (Reprinted from *Trends Microbiol.,* 4, 394-400, 1996. Slifka, M.K. and Ahmed, R., Long-term humoral immunity against viruses: revisiting the issue of plasma cell longevity. © 1996, with permission of Elsevier Science.)

CELL-MEDIATED IMMUNITY

This is perhaps the most important defense against viral infections and results from specific and nonspecific events. Viruses from infected cells induce lymphocytes (or other cells) to produce interferon and to destroy infected cells. Destruction of infected cells is carried out by cytotoxic T lymphocytes (CTLs). Lymphocytes also secrete chemotactic factors to recruit macrophages and to prevent them from migrating away (MIF). Persistent viruses often have evolved mechanisms to escape CTL recognition. One strategy, followed by adenoviruses and herpes simplex viruses, is to subvert antigen presentation by MHC class I molecules.

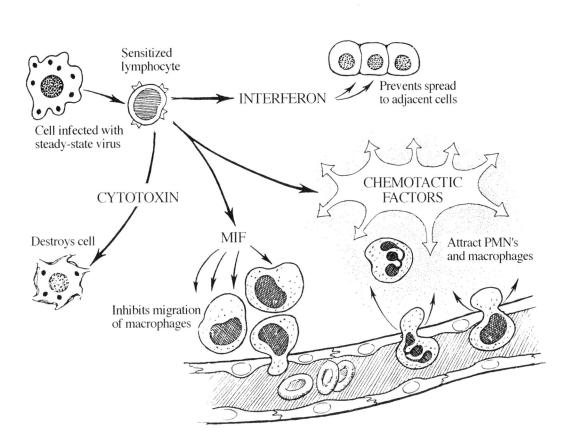

FIGURE 61

Cell-mediated events in viral infections. The interaction of the virus with sensitized lymphocytes releases cytotoxin which destroys the infected cells and generates chemotactic factors which recruit macrophages to remove debris and, in addition, may release interferon to prevent a spread of viruses to adjacent cells. Simultaneously, macrophages are activated and their migration is inhibited. MIF, migration inhibition factor. (From Bellanti, J.A, *Immunology*, W.B. Saunders, Philadelphia, 1971, 286. With permission.)

The players in cell-mediated immunity are specific (T lymphocytes) or nonspecific (NK cells, macrophages). T lymphocytes include three functional groups: (1) cytotoxic T cells which kill target cells, (2) suppressor cells which suppress immune reactions, and (3) helper cells which produce cytokines that stimulate B lymphocytes. T cells are major producers of interferon. They include two groups with distinct surface markers, CD8 and CD4. CD8+ cells are essentially cytotoxic and CD4+ cells are helpers. Cytotoxic T lymphocytes lyse infected cells by release of perforin, a protein that creates holes in the cytoplasmic membrane of the target cell. Lesions resemble complement-mediated lysis.

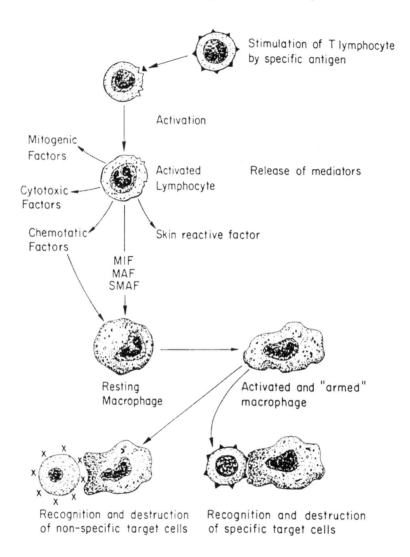

FIGURE 62

T lymphocyte-macrophage effector mechanisms. (From Alexander, J.W. and Good, R.A., *Fundamentals of Clinical Immunology,* W.B. Saunders, Philadelphia, 1977, 103. With permission)

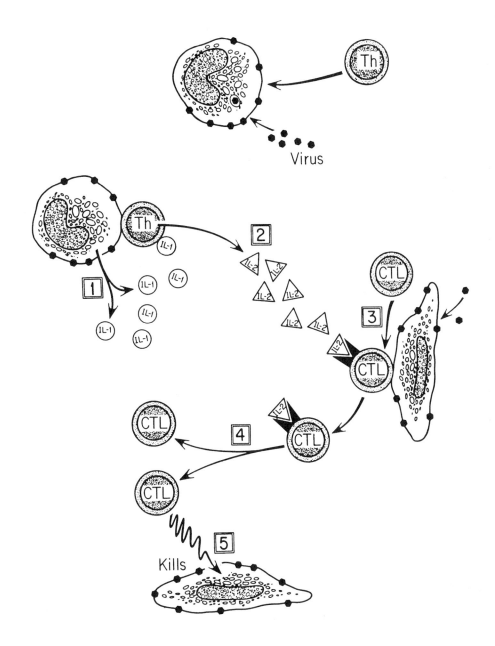

FIGURE 63

Some events leading to the generation and activation of cytotoxic T cells (CTL). (1) A T helper cell recognizes viral antigens in the macrophage; the macrophage releases IL-1. (2) The activated T cell then releases IL-2. (3) A CTL establishes contact with viral antigens displayed on a fibroblast and acquires receptors for IL-2. (4) After binding to IL-2, the CTL proliferates. (5) The expanded clone of CTL establishes contact with the virally infected fibroblast and kills it. (From Unanue, E.R. and Benacerraf, B., *Textbook of Immunology*, 2nd ed., Williams & Wilkins, Baltimore, 1984, 162. © Lippincott Williams & Wilkins. With permission.)

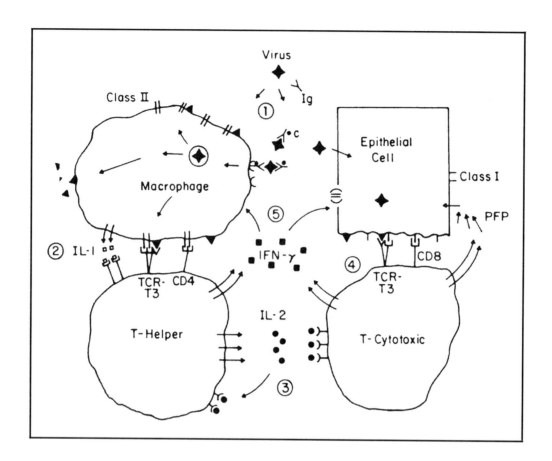

FIGURE 64

Host interactions with virus leading to cellular immune responses. 1. Virus is processed by a phagocyte (e.g., macrophage) and viral antigens are presented on antigen-presenting cells (APC, e.g., macrophage, dendritic cell). This process is enhanced by macrophage Fc and complement receptor binding of immunoglobulin (Ig) and complement (C) components bound to the virus. 2. The antigen receptors (TCR-T3 complex) of T-helper (T_H) cells recognize antigen and self-class II MHC, whereas CD4 molecules enhance the binding interactions of T_H with common class II MHC determinants. Interleukin-1 (IL-1) from the APC promotes T_H cell stimulation by antigen. 3. Interleukin-2 (IL-2) from activated T_H enhances T cytotoxic (T_C) and T_H activity. 4. Activated T_C cells recognize viral antigen and self-class I MHC, whereas CD8 molecules enhance the binding interactions with common class I determinants. (5) Activated T cells produce IFN-γ which stimulates macrophages and induces increased class II MHC expession on AC and target cells. Activated T_C cells also produce pore-forming protein (PFP, perforin), which can lyse infected targets. (From Sanfilippo, A., Balber, A.E., Granger, D.I., and McKinney, R.E., in *Zinsser Microbiology,* 19th ed., Joklik, W.K., Willett, H.P., Amos, D.B., and Wilfert, C.M., Eds., Appleton & Lange, Norwalk, 1988, 277. © 1988 McGraw-Hill Companies. With permission.)

NK or "natural killer" cells are large lymphocytes in the peripheral blood that have no immunological memory and kill virus-infected cells. They are thought to have a role in innate immunity and immune surveillance and are the principal mediators of antibody-dependent cell-mediated cytotoxicity (ADCC, FIG. 58). The NK response peaks about 3 days after infection. Together with interferons, NK cells represent the first line of defense against viral infections.

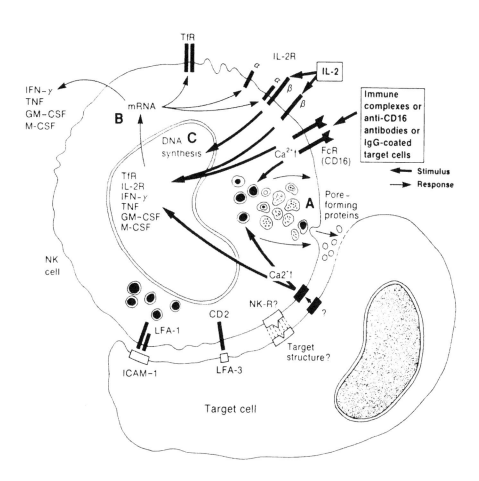

FIGURE 65

Model of NK cell activation following interaction with target cells or immune complexes. The interaction of NK cells with target cells involves unknown receptor(s) responsible for binding and NK cell activation (signal transduction). The latter involves enhanced phosphoinositide turnover and increase of intracellular Ca^{2+}, due to release of Ca^{2+} from intracellular stores and influx of extracellular Ca^{2+}, observed upon interaction of NK cells with NK cell-sensitive target cells. Similar signal transduction mechanisms are activated during interaction of CD16 FcR with ligands, i.e., with anti-CD16 antibodies, immune complexes, of IgG-coated target cells. Three types of response are observed: (A) activation of the cytotoxic mechanism, with morphological alteration and secretion of the content of granules, including cytotoxic molecules such as pore-forming proteins; (B) transcription of lymphokine and cell surface receptor genes and expression of their products, with a synergistic induction mediated by IL-2; (C) proliferation of NK cells, mostly induced by IL-2 interaction with the high-affinity IL-2 receptor (IL-2R) (a dimer with α and β chains) or with its β chain, but modulated by the regulatory effect of NK-target cell or FcR (CD16)-ligand interactions on the expression of the IL-2 receptor gene. (From Trinchieri, G., *Adv. Immunol.*, 47, 187, 1989. With permission.)

In contrast to B lymphocytes, T cells react to antigen presented by either of two classes of the MHC (major histocompatibility complex) pathway. In class I, endogenous antigen, usually made within the infected cell, is processed and presented to CD8[+] cells. The pathway is universal and found in most somatic cells. It is privileged by viruses because of their intracellular replication. In class II, extracellular antigen is taken in, processed, and presented to CD4[+] cells.

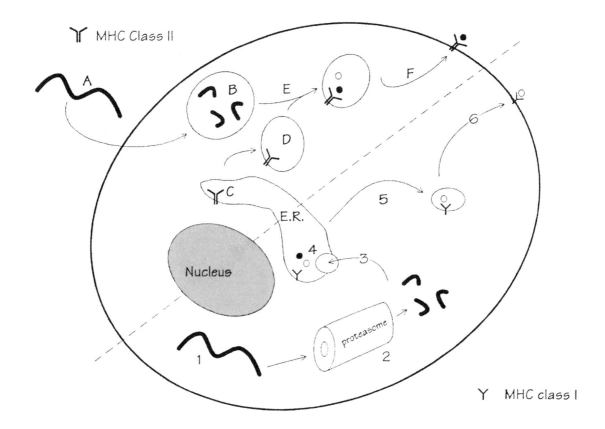

FIGURE 66

Antigen processing in MHC class I and class II pathways. In the class I pathway *(lower half),* newly synthetized intracellular virus protein (1) is degraded to shorter peptides (2), perhaps in an organelle named the proteasome. The fragments enter the endoplasmic reticulum (ER) (3). One fragment (white circle) is selected by the MHC class I molecule (4), the complex passes into the Golgi apparatus (5) and finally reaches the cell surface. In the class II pathway *(upper half),* extracellular exogenous virus protein (A) is endocytosed and degraded in an acidic compartment (B). The class II molecule is synthetized within the ER (C) and is transported through the Golgi network (D). Vacuole B and endosome D fuse and a viral peptide is bound to the class II molecule (E). The mature complex reaches the cell membrane (F). (Author's legend) (From Whitton, J.L. and Oldstone, M.B.A., in *Fields Virology,* 3rd ed., Vol. 1, Fields, B.N., Knipe, D.M., and Howley, P.M., Eds.-in-chief, Lippincott-Raven, Philadelphia, 1996, 345-374. © Lippincott Williams & Wilkins. With permission.)

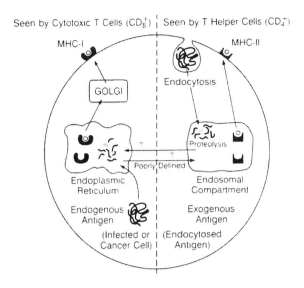

FIGURE 67

Two faces of the immune system: antigen processing and presentation. MHC, major histocompatibility complex. (From Hilleman, M.R., *Progr. Med. Virol.,* 39, 1, 1992. With permission.)

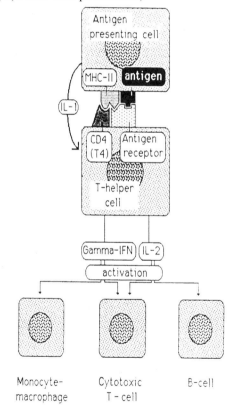

FIGURE 68

Central regulatory mechanisms of immune responses. Antigen is processed by antigen-presenting cells and presented to T4-cells in conjunction with a MHC II molecule. T4 cells recognize the antigen/MHC II complex by the antigen receptor. Additional requirements for T4 cell activation include interleukin 1 (IL-1) and the interaction of the CD4 molecule with the nonpolymorphic part of MHC II. Immune effector systems (monocytes/macrophages, cytotoxic T cells, B cells, natural killer cells) require T4 assistance which is provided by the release of lymphokines such as gamma interferon (IFN-_) and interleukin 2 (IL-2). (Fig. 14 from Schüpbach, J., *Human Retrovirology, Facts and Concepts, Curr. Topics Microbiol. Immunol.,* 142, 52, 1989. © Springer-Verlag. With permission.)

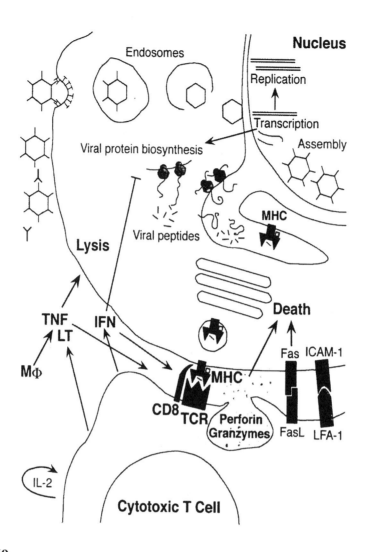

FIGURE 69

Role of major histocompatibility complex (MHC) molecules and cytotoxic T cells (CTLs) for clearing virus-infected cells. Adenoviruses are taken as an example. Complete virions accumulate in the nucleus and only a small fraction of viruses are released while the cell is still alive. Viral peptides are bound inside the cell by MHC class I molecules and are transported to the cell surface where they are recognized by the T cell receptor (TCR) on the surface of CTLs. Activation of CTLs induces the release of perforin and granzymes, which promote lysis of the infected cell. A second mechanism of CTL-mediated cell death is induction of apoptosis by interaction of the Fas ligand (FasL) on the T cell with the Fas/Apo-1 receptor on the target cell surface. CTLs also secrete tumor necrosis factor (TNF) and lymphotoxin (LT), which might be involved in target cell killing. However, most TNF is produced by macrophages (Mϕ) and monocytes during the early phase of the immune response. Another factor secreted by CTLs is interferon γ(IFN-γ), whose multiple effects include inhibition of viral protein synthesis, activation of macrophages and stimulation of MHC expression. Also, TNF can upregulate MHC expression on the cell surface. Therefore, both cytokines promote antigen presentation and T cell recognition (arrows). ICAM-1, intercellular adhesion molecule 1; IL-2, interleukin 2; LFA-1, lymphocyte function-associated antigen 1. (Reprinted from *Trends Microbiol.,* 4, 107-112, 1996. Burgert, H.-G., Subversion of the MHC class I antigen-presentation pathway by adenoviruses and herpes simplex virus. © 1996, with permission of Elsevier Science.)

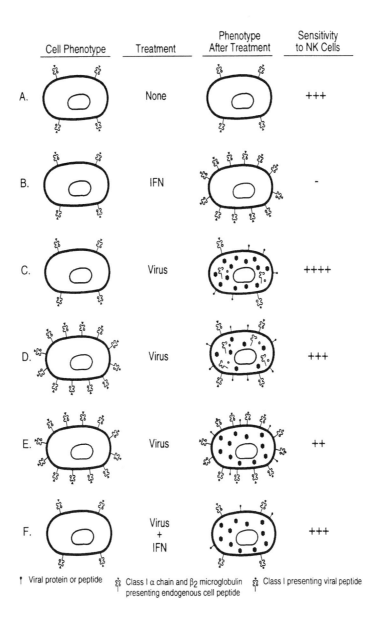

FIGURE 70

Alterations in target cell class I antigens and sensitivity to NK cells by IFN and viral infections. (A) Target cells expressing low levels of class I MHC antigens are susceptible to NK cells, and the IFN induced by virus infections can upregulate class I expression (B) and thereby protect target cells from cell-mediated lysis. Viral infections, through a variety of mechanisms, can downregulate class I expression and may through this mechanism render cells more susceptible to NK cells (C and D). It is possible though not yet definitely shown that viruses may interfere with the class I molecule-induced negative signal by the replacement of the presented cellular peptides with virus-encoded (or induced) peptides (E). Cells already infected with viruses may become resistant to IFN and may not upregulate class I antigens, thereby remaining sensitive to NK cells (F). (From Brutkiewicz, R.R. and Welsh, R.M., *J. Virol.,* 69, 3967, 1995. With permission.)

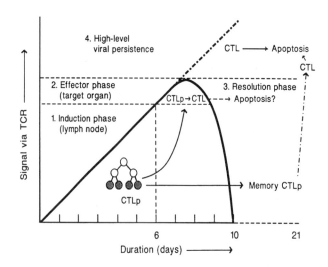

FIGURE 71

A model of the development of CD8$^+$ T cell response in virus infections. The central concept is that the extent of signaling via the T cell receptor (TCR) determines the fate of CD8$^+$ cells developing from T lymphocyte precursors (CTLp). Clonal expansion and differentiation of virus-specific CD8$^+$ CTLp occurs (stage 1) in the regional lymph node (RLN). The activated CTLp then exit the RLN and travel via the blood to the target organ, where exposure to large numbers of infected cells leads to the development of CTL effector function (stage 2). Some of these CTL may undergo apoptosis during either the effector or the resolution (stage 3) phase of the inflammatory process, after virus is cleared. Failure to eliminate the virus (stage 4) may lead to persistent stimulation and eventual exhaustion of the CTL clone. With most viruses (though not LCMV) the eventual consequence would be death from virus-induced cytopathology. Whether the stimulation of CTLp in the lymph node (stage 1) is indeed continuous, or is better represented as an early encounter with antigen that is not repeated until the T cells relocate to the target organ (stage 2), is yet to be resolved. Clonal deletion in the thymus may reflect that the TCR-mediated signal postulated for stage 1 leads either to anergy or to apoptosis for developing lymphocytes. (Reprinted from *Trends Microbiol.,* 1, 207-209, 1993. Doherty, P.C., Immune exhaustion: driving virus-specific CD8$^+$ T cells to death. © 1993, with permission from Elsevier Science.)

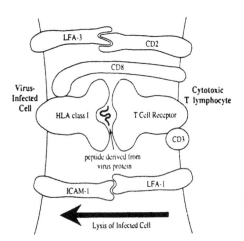

FIGURE 72

CD8-positive cytotoxic T lymphocyte recognition of a virus-infected cell. A short peptide derived from a virus protein is presented to the T cell receptor in association with a particular HLA class I molecule. The CTL: target-cell interaction is stabilized by CD8 binding to the HLA class I molecule and by the specific binding of various adhesion molecules (only the LFA-1/ICAM-1 and LFA-3/CD2 interactions are shown). (From Lee, S.P., *Sem. Virol.,* 5, 281, 1994. With permission.)

Both direct immunologic tissue damage and autoimmunity are seen in cell-mediated injury. The former is seen in adult mice infected with lymphocytic choriomeningitis (LCM) virus and caused by cytotoxic lymphocytes. Damage may be prevented by immunosuppression, for example by X-ray irradiation or anti-lymphocyte serum. Autoimmunity is seen in murine hepatitis infection of the CNS, in which the myelin sheath of nerve axons becomes a target for T cells.

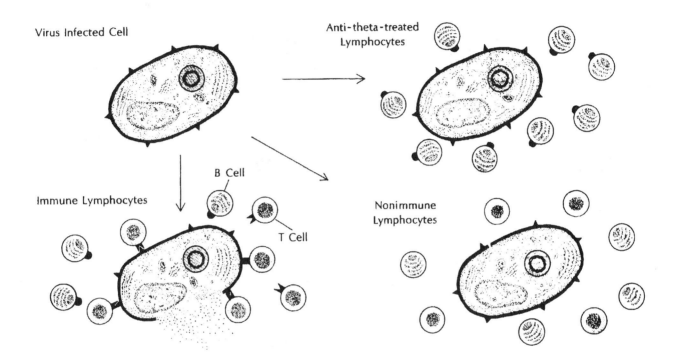

FIGURE 73

Role of cell-mediated immunity in immunologic injury as shown in experiments with LCM virus. The virus alone does not destroy the cell it infects, but in the presence of immune lymphocytes, cell lysis occurs. Evidence for T-lymphocyte involvement comes from incubating the immune lymphocytes with anti-theta antibody, which destroys T but not B lymphocytes. After incubation, cell lysis is obviated. The diagram is not meant to imply immune lymphocytes may not interact with infected cells, but only that any such interaction is usually not cytolytic. Lymphocytes from nonimmune animals do not lyse infected cells. (Reproduced with permission. From Notkins, A.I., Viral infections: mechanisms of immunologic defense and injury. *Hospital Practice,* 9 (9), 65. © 1974 The McGraw-Hill Companies, Inc. Illustration by Nancy Lou Riccio.)

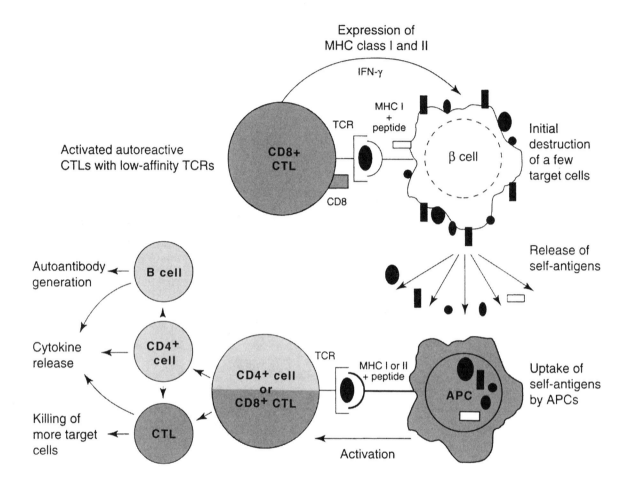

FIGURE 74

Events involved in virus-induced autoimmune diseases. Peripheral unresponsiveness is broken through the activation of (cross)reactive (probably low-affinity) T cells. These cells home to the target organ, where self-antigen is recognized. In the absence of co-stimulatory signals, anergy results without tissue (cell) destruction. In the presence of the required activation signals, autoreactive T cells are activated. Damage can occur either indirectly through the release of cytokines or directly through cytotoxic-T-cell-mediated (CTL-mediated) lysis of the target cells (such as β cells). Once damage has occurred, a chain reaction follows that gradually leads to clinically overt autoimmune disease. APC, antigen-presenting cell; IFN- γ, interferon γ; MHC, major histocompatibility complex; TCR, T cell receptor. (Reprinted from *Trends Microbiol.,* 3, 424-430, 1995. Von Herrath, M.G. and Oldstone, M.B.A., Role of viruses in the loss of tolerance to self-antigens and in auto-immune diseases. © 1995, with permission of Elsevier Science.)

Chapter 6

VERTEBRATE VIRUSES BY FAMILY

6.I. ADENOVIRIDAE

Linear dsDNA
Cubic, naked

The name is derived from adenoid tissue, from which the first representatives were isolated. Adenoviruses comprise the genera *Mastadenovirus* (mammals) and *Aviadenovirus* (birds). Particles are icosahedra of 80-90 nm in diameter. Viruses are generally host-specific. Human adenoviruses are associated with benign respiratory disease, conjunctivitis, and gastroenteritis; human adenovirus type 11 is strongly linked to hemorrhagic cystitis. Although there is no association with human cancer, serotypes 12, 18, and 31 are highly oncogenic when injected into baby rodents. Canine adenoviruses cause hepatitis and respiratory disease. Avian adenoviruses are noted for causing hemorrhagic enteritis and pulmonary edema. Adenoviruses are currently investigated for their regulation of cell apoptosis (below); diagrams illustrating adenovirus pathogenesis seem to be nonexistent. Adenovirus infection and adenoviral gene E1A stimulate cellular replication. This results in suicidal elimination of virus-infected cells by apoptosis. To counter premature cell death, adenoviruses have evolved a mechanism encoded by gene E1B. The 55K protein of E1B interferes with the p53 tumor suppressor protein and the E1B 19K protein inhibits p53-dependent apoptosis.

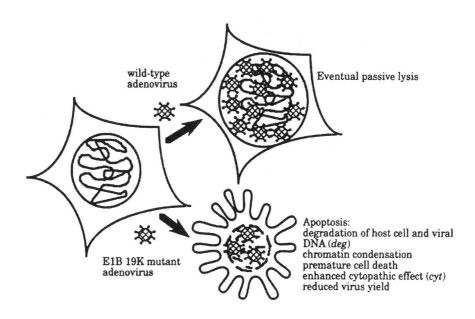

FIGURE 75

Regulation of apoptosis in adenovirus-infected cells. Infection with wild-type adenovirus *(top)* results in suppression of apoptosis by E1B 19K protein. Infection with an E1B 19K mutant *(bottom)* leads to apoptosis. (From White, E., *Sem. Virol.,* 5, 341, 1994. With permission.)

6.II. ARENAVIRIDAE

Linear (-) sense ssRNA, segmented
Helical, segmented

This small family consists of a single genus of mostly rodent-pathogenic viruses. Particles are enveloped, roughly spherical, and usually 110-130 nm in diameter. Most viruses induce a persistent, frequently asymptomatic disease in mice. Humans are infected by contact with rodents (and humans) and contract hemorrhagic fevers, the most important of which is the highly lethal Lassa fever (West Africa). Symptoms include hemorrhagic rash, myalgia, and shock. Lymphocytic choriomeningitis virus (LCMV) of mice is a laboratory model for the study of persistent viral infections. Intracerebral inoculation of previously unexposed adult mice induces inflammatory changes and death, but infection of immunosuppressed adult mice, fetuses, or newborns leads to persistent, inapparent infection. LCMV is contracted *in utero* and maintained as a life-long, inapparent infection. The virus occasionally induces in humans an influenza-like syndrome.

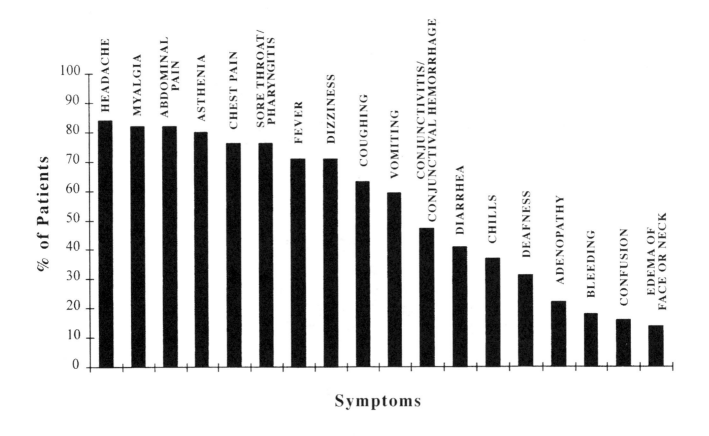

FIGURE 76

Symptoms observed in 51 cases of Lassa fever in Sierra Leone in 1996. (Adapted from Bausch, D.G. and Rollin, P.E., *Ann. Inst. Pasteur/Actualités*, 8, 223, 1997. With permission.)

LYMPHOCYTIC CHORIOMENINGITIS

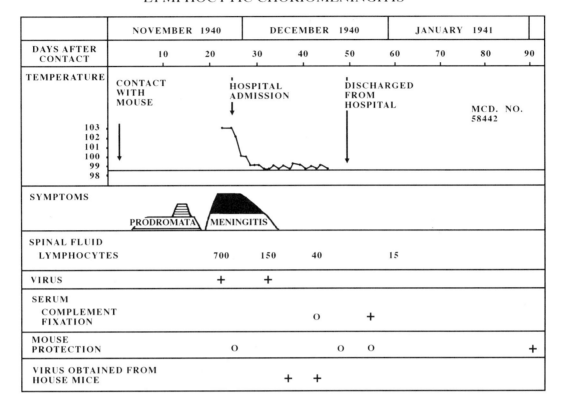

FIGURE 77

Course and essential laboratory data in a case of spontaneous lymphocytic choriomeningitis, meningeal form. (From Farmer, T.W. and Janeway, C.A., *Medicine,* 21, 1-64, 1942. © Lippincott Williams & Wilkins. With permission.)

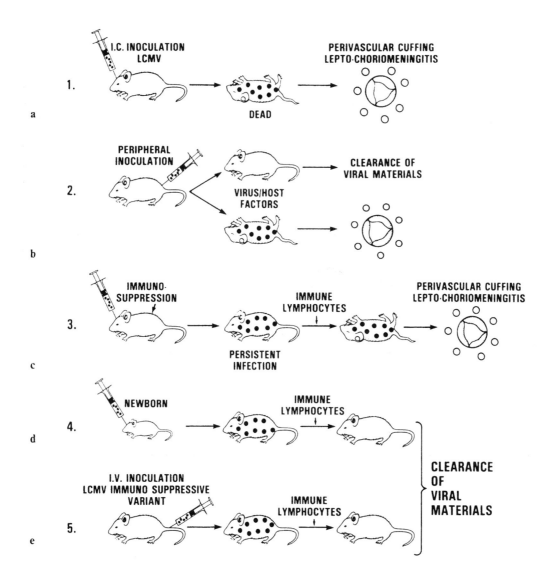

FIGURE 78

Experimental model for establishing acute and persistent LCMV infection. a. Intracerebral inoculation of adult immunocompetent animals with sufficient amounts of wild-type LCMV leads to death within 6-10 days associated with perivascular cuffing and infiltration into the lymphomeninges, choroid plexus, and ventricles. b. Peripheral inoculation of adult immunocompetent animals, dependent on route of inoculation and dose, genetic background of the host, and other factors leads either to clearance of viral materials and immunity or death with a histopathologic picture similar to that of mice dying after intracerebral inoculation. c. Adult immunocompetent mice inoculated intracerebrally with an ordinarily lethal dose of virus and receiving immunity suppression by a variety of means (irradiation, thymectomy, antithymocytic or lymphocytic sera, cytoxan, etc.) develop a persistent infection. Virus is found primarily in the leptomeninges and ependymal cells lining the ventricles. Transfer of LCMV-immune lymphocytes rapidly causes death. d. In contrast, inoculation of newborn mice with most wild-type strains of LCMV or e. inoculation of adult immunocompetent mice with immunosuppressive variants of LCMV leads to persistent infection with virus expressed in most tissues of the body and found primarily in neurons of the central nervous system. Transfer of virus-specific immune lymphocytes promotes clearance of the virus. (Fig. 1 from Oldstone, M.B.A., in *Arenaviruses. Biology and Immunotherapy,* Oldstone, M.B.A., Ed., *Curr. Topics Microbiol. Immunol.,* 134, 211, 1987. © Springer-Verlag. With permission.)

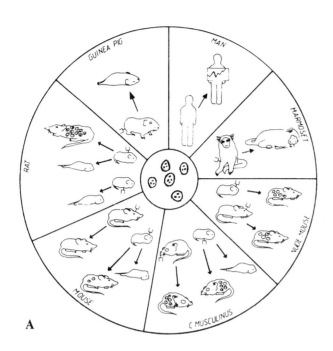

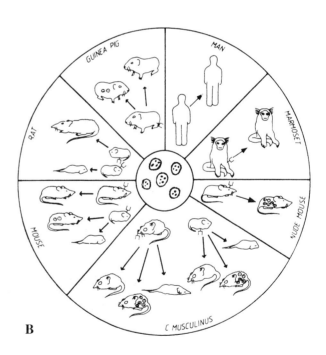

FIGURE 79

Patterns of evolution of infection in man and other species by the pathogenic XJ (A) and the attenuated XJCl3 (B) strains of Junín virus. Outcome of disease and its eventual persistence depend on numerous factors, e.g., virus strain, inoculation route, age at the moment of infection, immunological status, and animal species. (Adapted from Fig. 1 from Weissenbacher, M.C., Laguens, R.P., and Coto, E.E., in *Arenaviruses. Biology and Immunotherapy,* Oldstone, M.B.A., Ed., *Curr. Topics Microbiol. Immunol.,* 134, 79, 1987. © Springer-Verlag. With permission.)

6.III. BUNYAVIRIDAE

Linear (+) sense ssRNA, segmented
Helical, enveloped

This large family is named after Bunyamwera, a locality in Uganda. It includes four genera of vertebrate viruses, most of which are transmitted by arthropods in which they multiply, and one genus of plant viruses *(Tospovirus)*. Particles are enveloped, roughly spherical, and 80-120 nm in diameter. Most infections are asymptomatic. Clinical infections present general symptoms such as fever, myalgia, arthralgia, and rash, which may be followed by bleeding into the skin or from body orifices and hypotensive shock, pulmonary edema, or encephalitis. The most notable diseases are Korean hemorrhagic fever, Rift Valley fever, and hantavirus pulmonary syndrome (U.S.A.). The tick-borne Uukuniemi virus, a member of the *Phlebovirus* genus, induces lethal encephalitis in mice and is an important experimental model.

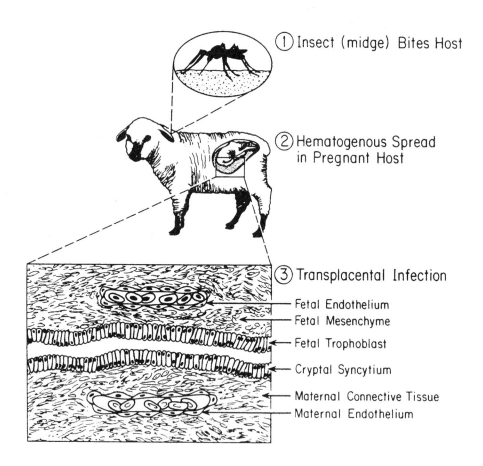

FIGURE 80

Stages involved in infection of the fetus by a bunyavirus. (1) The insect carrying the virus bites a susceptible host and causes virus replication at or near the primary site of inoculation. (2) The resulting short-term viremia results in virus spreading within the reticuloendothelial system in a pregnant host. (3) Viruses must then replicate successfully in the six tissue layers between maternal and the fetal circulation resulting in infection of the fetus. (From Parsonson, I.M. and McPhee, D.A., *Adv. Virus Res.,* 30, 279, 1985. With permission.)

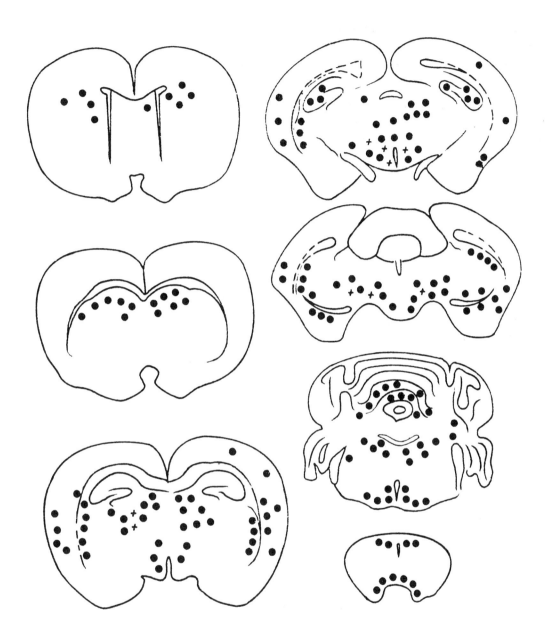

FIGURE 81

Distribution of inflammatory lesions in the brain of suckling *Microtus arvalis* 5-7 days after intracerebral inoculation with Uukuniemi virus. Dots, inflammatory infiltrates; +, neuronophagia. (From Kozuch, O., Rajcáni, J., Sekeyová, M., and Nosek, J., *Acta Virol.*, 14, 163, 1970. With permission.)

6.IV. CALICIVIRIDAE

Linear (+) sense ssRNA, nonsegmented
Cubic, naked

These viruses occur in a wide range of vertebrates and are frequently associated with gastroenteritis. Particles are icosahedra of 35-39 nm in diameter and characterized by cup-shaped depressions (Latin, *calix,* "cup"). Hepatitis E virus (HEV) has the general characteristics of caliciviruses and is considered by many as a member of this family, but has recently been reclassified as a "floating" unassigned genus. Its epidemiology is similar to that of hepatitis A virus. HEV is transmitted by water and food and has caused explosive epidemics especially in India. Symptoms are similar to those of other types of hepatitis (see *Hepadnaviridae).* Mortality of pregnant women is particularly high. As in hepatitis A, there is no tendency toward chronicity.

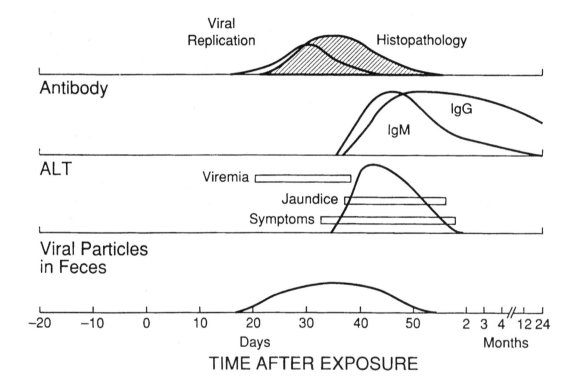

FIGURE 82

Clinical and serologic events in a typical case of acute E hepatitis. (Modified from reference 96, from Purcell, R.H., in *Fields Virology,* 3rd ed., Vol. 2, Fields, B.N., Knipe, D.M., and Howley, P.M., Eds.-in-chief, Lippincott-Raven, Philadelphia, 1996, 2831-2843. © American Public Health Association and Lippincott Williams & Wilkins. With permission.)

6.V. CORONAVIRIDAE

Linear (+) sense RNA, nonsegmented
Helical, enveloped

The family has two genera. Particles are spherical to pleomorphic, usually 160-180 nm in diameter, and characterized by a "crown" of large, club-shaped projections. Members of the *Torovirus* genus may be disk-, rod-, or kidney-shaped. Viruses infect a wide range of mammals and birds, causing respiratory, enteric, or neurologic disease, some of which are economically important. Human coronaviruses cause minor respiratory disease including common cold. Mouse hepatitis is a useful experimental model for the study of persistent infections and neural spread.

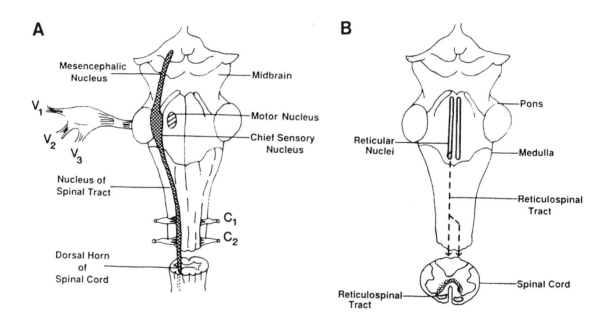

FIGURE 83

After intranasal inoculation, mouse hepatitis virus (MHV) gains entry into the central nervous system via the olfactory nerves. Some mice develop demyelinating encephalomyelitis several weeks later, with virus always present in the spinal cord. (A) MHV seems to reach the spinal cord via well-defined neuroanatomical pathways. The shaded area indicates the sensory nuclei of the trigeminal nerve. Continuity of the nucleus of the spinal tract of the trigeminal nerve with the dorsal horn *(substantia gelatinosa)* of the upper cord is shown. C_1, C_2, first and second cervical spinal segments. V_1, V_2, V_3, first, second, and third division of trigeminal nerve. (B) After entry via the olfactory nerves, virus may also pass to reticular nuclei in the pons and the medulla and from there enter the spinal cord via the reticulospinal tract. Reticulospinal fibers terminate on cells in the anterior portion of the spinal cord, indicated by dots in the cross section. (Author's note: In the original publication, MHV was thought to enter via olfactory and trigeminal nerves. The exclusive use of the olfactory route was later established; see Barnett, E.M. and Perlman, S., *Virology,* 194, 185, 1993) (From Perlman, S., Jacobsen, G., Olson, A.L., and Afifi, A., *Virology,* 175, 418, 1990. With permission.)

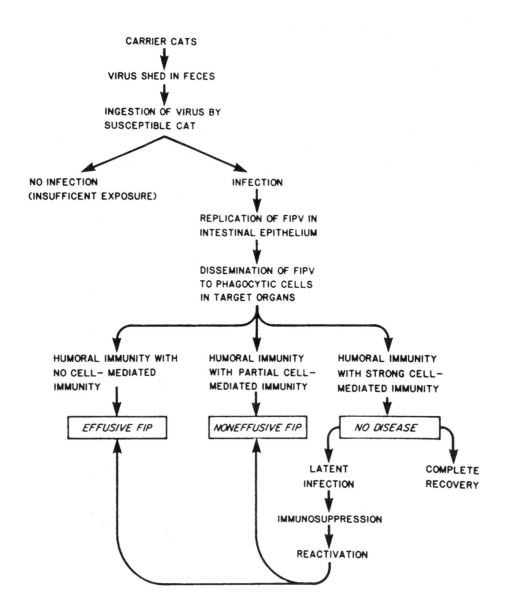

FIGURE 84

Possible pathogenesis of feline infectious peritonitis. FIP, feline infectious peritonitis virus; FIPV, FIP virus. (From Pedersen, N.C., Black, J.W., Boyle, J.F., Evermann, J.F., McKeirnan, A.J., and Ott, R.L., in *Molecular Biology and Pathogenesis of Coronaviruses,* Rottier, P.J.M., Van der Zeijst, B.A.M., Spaan, W.J.M., and Horzinek, M.C., Eds., Plenum Press, New York, 1983, 365. With permission.)

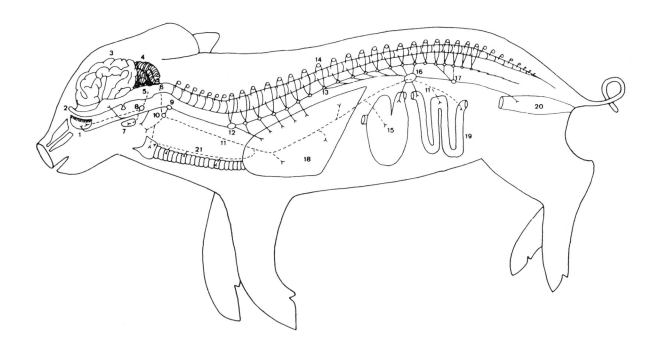

FIGURE 85

Tissues, nerve paths, and ganglia involved into the pathogenesis of Vomiting and Wasting disease in swine. 1. Nasal mucosa. 2. Olfactory bulb. 3. Cerebrum. 4. Cerebellum. 5. Pons varoli. 6. Medulla oblongata. 7. Tonsils. 8. Trigeminal ganglion. 9. Cranial cervical ganglion. 10. Inferior vagal ganglion. 11. Vagal nerve. 12. Stellate ganglion. 13. Sympathic trunc. 14. Dorsal root ganglia. 15. Stomach. 16. Solar ganglion. 17. Caudal mesenteric ganglion. 18. Lungs. 19. Small intestine. 20. Rectum. 21. Recurrent vagal nerve. (From Andries, K. and Pensaert, M., in *Biochemistry and Biology of Coronaviruses*, Ter Meulen, V., Siddell, S., and Wege, H., Eds., Plenum Press, New York, 1981, 399. With permission.)

6.VI. FILOVIRIDAE

Linear (-) sense ssRNA, nonsegmented
Enveloped, helical

This very small family, a member of the order *Mononegavirales*, comprises only two genera, represented by Marburg and Ebola viruses. Both cause extremely dangerous hemorrhagic fevers in man. Ebola subtype Reston is pathogenic for monkeys only. Virions measure about 800-1000 x 80 nm, have an inner component resembling that of rhabdoviruses, and are very pleomorphic (filamentous, branched, circular, U-shaped). The human disease is a severe pantropic infection that involves nearly all systems and is characterized by generalized bleeding and shock (up to 88% mortality). Macrophages, fibroblasts, and epithelial cells may be the preferred sites of viral replication in early stages. Increased levels of the cytokine TNF-_ suggest that mediator-induced vascular permeability is essential to the shock syndrome. The bleeding tendency could be explained by direct endothelial damage and indirectly by cytokine-mediated processes.

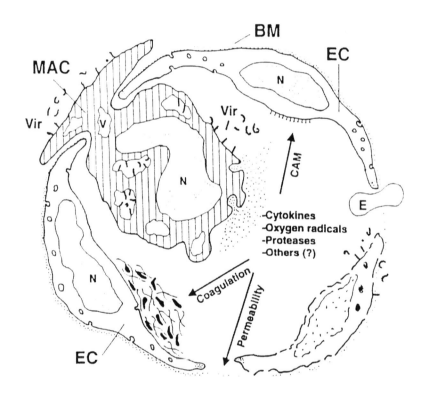

FIGURE 86

Possible role of macrophages and endothelial cells in evolution of hemorhagic filovirus fevers. BM, basal membrane; CAM, cellular adhesion molecule; E, erythrocyte; EC, endothelial cell; MAC, macrophage; N, nucleus; V, vacuole; Vir, viral particle. In cooperation with H.J. Schnittler, Physiological Institute, Munster, Germany. (From Feldmann, H., Volchkov, V.E., and Klenk, H.-D., *Ann. Inst. Pasteur/Actualités,* 8, 207, 1997. With permission.)

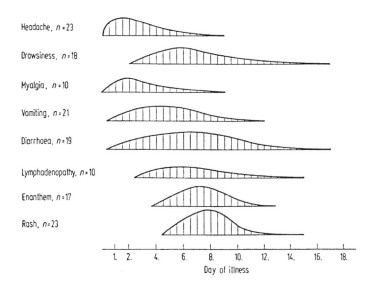

FIGURE 87

Marburg virus disease. Incidence and duration of main signs and symptoms: height of curves in top part is determined by the number of patients with the listed symptoms on a given day. (Upper part of Fig. 3 from Martini, G.A., in *Marburg Virus Disease,* Martini, G.A. and Siegert, R., Eds., Springer, New York, 1971, 1. © Springer-Verlag. With permission.)

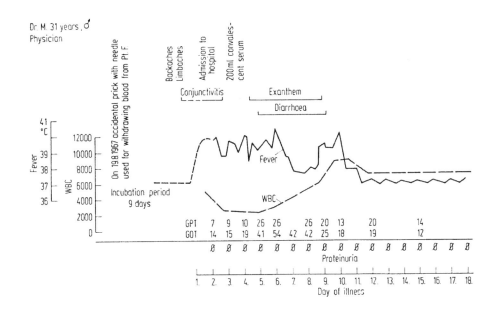

FIGURE 88

Course of illness in a contact case, a medical doctor who was treated with convalescent serum and survived. (The former patient became a well-known professor of internal medicine; W. Stille, personal communication) (Fig. 8 from Stille, W. and Böhle, E., in *Marburg Virus Disease*, Martini, G.A. and Siegert, R., Eds., Springer, New York, 1971, 10. © Springer-Verlag. With permission.)

6.VII. FLAVIVIRIDAE

Linear (+) sense RNA, nonsegmented
Cubic, enveloped

This family has three genera, *Flavivirus* (group B arboviruses including the agents of yellow fever and dengue), *Pestivirus* (animal pathogens including the virus of bovine diarrhea virus), and *Hepacivirus* (human hepatitis C). Particles are spherical, 40-60 nm in diameter, and consist of an envelope, an icosahedral capsid, and RNA. Flaviviruses were formerly classified as members of the *Togaviridae* family. Most members of the large *Flavivirus* genus multiply in arthropods (mosquitoes, ticks) and are transmitted by these to vertebrates (mammals and birds).

In humans, flaviviruses frequently cause fever with arthralgia and meningoencephalitis. Yellow fever, the most important and lethal human arbovirus disease, is characterized by hepatitis, nephritis, and extensive hemorrhages, especially in the intestinal tract. Although many infections are inapparent, yellow fever has had an enormous negative effect on the development of tropical countries. Dengue is an unpleasant infection with severe headache and pains in the back, muscles, and joints ("break-bone fever"). Dengue hemorrhagic fever, a very serious form characterized by profuse bleeding into the skin and mucosae, is becoming increasingly frequent in many tropical and subtropical countries. Death occurs by hypotensive shock. Dengue hemorrhagic fever appears to be induced by immunopathological mechanisms mediated by cytokines (Fig. 93).

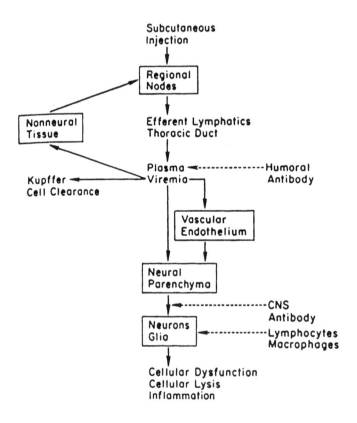

FIGURE 89

Sequential development of flavivirus infections. Boxes indicate sites of virus replication, and dotted arrows indicate immunologic defense mechanisms. (From Nathanson, N., in *St. Louis Encephalitis*, Monath, T.P., Ed., American Public Health Association, Washington, 1980, 201. © APHA, with permission.)

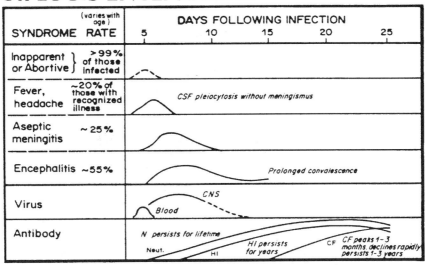

FIGURE 90

Clinical features of St. Louis encephalitis. (Reprinted with permission from Monath, T.P., in *Virology and Rickettsiology,* Vol. I, Part 2, Hsiung, G.-D and Green, R.H., Eds., CRC Press, Boca Raton, 1978, 261. Copyright CRC Press, Boca Raton, Florida.)

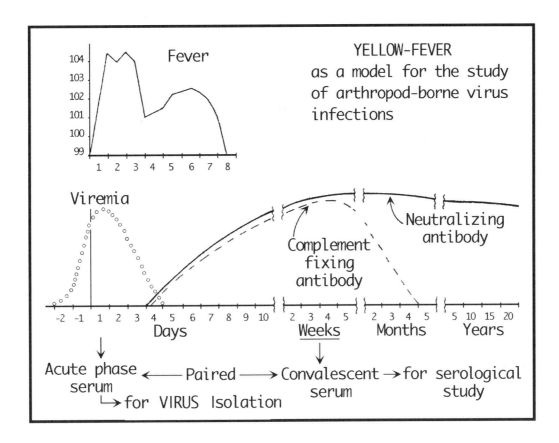

FIGURE 91

Time course of fever, viremia, and antibody in human infection with yellow fever virus. (From Kerr, J.A., *Indian J. Med. Sci.,* 7, 338, 1953. With permission.)

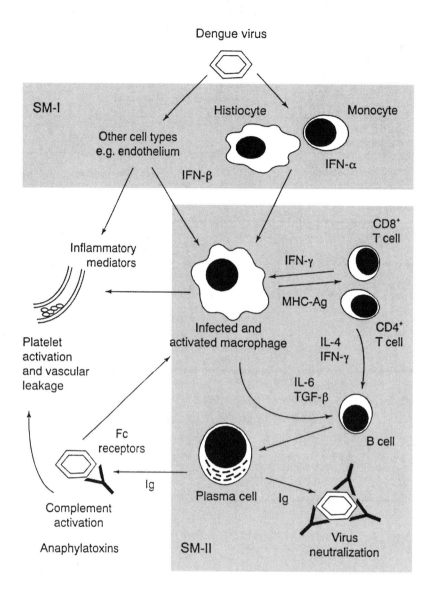

FIGURE 92

Immune response and inflammatory processes involved in dengue virus (DV) infections. Involvement of cytotoxic T cells (CD4+ or CD8+) has not been indicated because an *in vivo* role for this effector function in DV infections remains unknown. It is proposed that the main selective milieu (SM-1) for virus variants with increased virulence (more cytopathic and/or inflammogenic) exists or occurs prior to the activation of most of the virus-specific immune and nonspecific inflammatory reactions (particularly in young children). Inflammatory and immune reactions, such as virus-antibody complex formation and cytokine production by activated T cells, may in some cases enhance these *a priori* selective forces, creating selective milieu II (SM-II). Ag, antigen; IFN, interferon; Ig, immunoglobulin; IL, interleukin; MHC, major histocompatibility complex; TGF, transforming growth factor. (Reprinted from *Trends Microbiol.,* 5, 409-413, 1997. Bielefeldt-Ohmann, H., Pathogenesis of dengue virus diseases: missing pieces in the jigsaw. © 1997, with permission of Elsevier Science.)

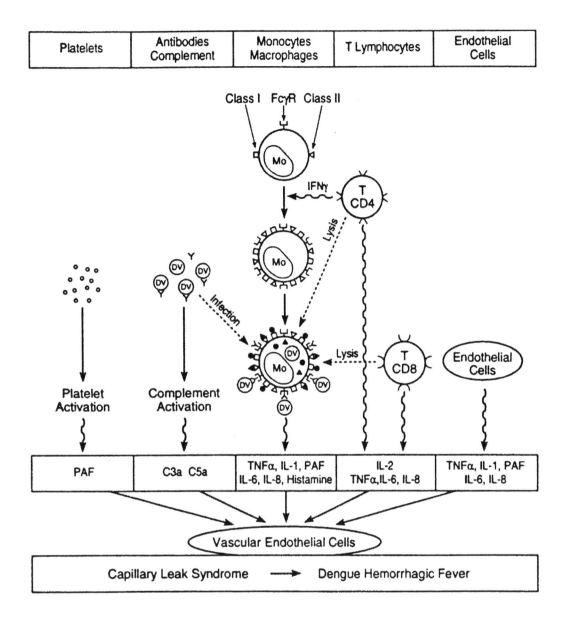

FIGURE 93

Possible role of cytokines and chemical mediators in dengue hemorrhagic fever. DV, dengue virus; IFN, interferon; IL, interleukin; Mo, monocyte; T, T lymphocyte; TFN, tumor necrosis factor. (From Kurane, L. and Ennis, F.A., *Sem. Virol.*, 5, 443, 1994. With permission.)

HEPATITIS C

This virus was the first to be identified without isolation of viral particles and cell cultures. Viral RNA was extracted from the serum of infected chimpanzees, transcribed into complemetary DNA (cDNA), and sequenced. The virus appears to be a member of the *Flaviviridae* family. The virus occurs world-wide. Its epidemiology and pathogenicity are very similar to that of hepatitis B virus (see below). The virus is essentially transmitted by blood and causes a wide spectrum of infections, ranging from inapparent disease to fulminant liver failure. As the virus of B hepatitis, it tends to cause chronic hepatitis, cirrhosis, and liver cancers. One mechanism of hepatitis C virus persistence may be the generation of immune-escape mutants.

[A]

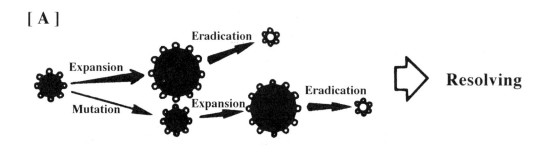

[B]

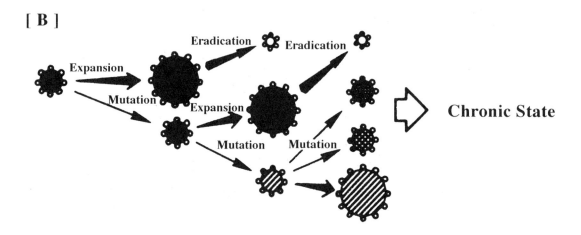

FIGURE 94

Quantitative and qualitative change of hepatitis C virus (HCV) with time in an infectant. Eradication of virus may be caused by neutralizing antibody, although this does not exclude the possibility that suppression of HCV proliferation is caused by immunological surveillance of the virus with cytotoxic T cells. (From Shimotohno, K., *Sem. Virol.,* 4, 305, 1993. With permission.)

6.VIII. HEPADNAVIRIDAE

Circular supercoiled dsDNA
Cubic, enveloped

This small family includes two genera of host-specific viruses, *Orthohepadnavirus* (mammals) and *Avihepadnavirus* (birds). Particles are about 45 nm in diameter and consist of an envelope, an icosahedral capsid, and one molecule of partially single-stranded DNA. Viruses replicate via reverse transcription using a virion-associated reverse transcriptase.

Human hepatitis B virus (HBV) has not been cultivated and is not cytopathogenic. In addition to HBV, the blood of infected individuals carries large quantities of particulate surplus envelope material, called hepatitis B surface antigen (HBsAg) or "Australia" antigen. Clinical symptoms of HBV infection result essentially from the destruction of virus-infected hepatocytes by cytotoxic T-lymphocytes. Human B hepatitis is an expanding disease and particularly frequent in East Asia and Africa. Asymptomatic carriers are the only reservoir. HBV is essentially transmitted by infected blood and sexual intercourse, via the placenta and perinatally; health-care professionals in contact with blood and intravenous drug users are at particular risk. The outcome of HBV infections is very variable and ranges from the asymptomatic carrier state to fulminant hepatitis and chronic forms evolving into cirrhosis and liver carcinoma. Typical acute disease is characterized by a long incubation period (45-120 days), anorexia, malaise, nausea, and vomiting. Jaundice appears after about one week and is accompanied by dark urine, pale feces, and high levels of serum bilirubin and transaminases. About 50% of cases are anicteric. HBV infections tend to evolve toward chronicity. Many viral and serologic markers are available to follow the evolution of the disease.

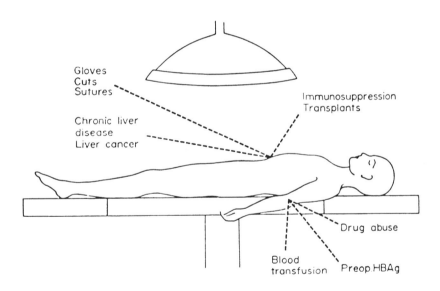

FIGURE 95

The hepatitis hazard for hospital workers and particularly surgeons comes from skin lacerations and pricks when handling blood from a patient with a positive hepatitis B antigen test, especially those with chronic liver disease or cancer, those on immuno-suppression or having organ transplantation or abusing drugs. Care in handling all blood specimens is emphasized. (From Sherlock, S., *Diseases of the Liver and Biliary System,* 6th ed., Blackwell Scientific Publications, Oxford, 1981, 256. With permission.)

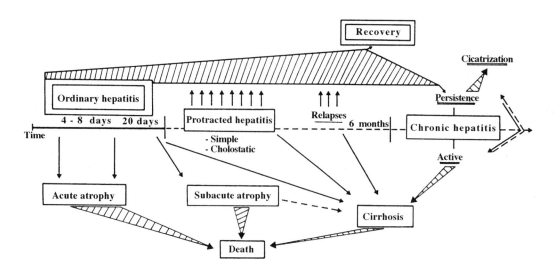

FIGURE 96

Evolution of B hepatitis. (Adapted from Garrigue, G., in *Virologie Médicale,* Maurin, J., Ed., Flammarion Médecine Sciences, Paris, 1985, 752. With permission.)

EVOLUTION OF HBV INFECTIONS

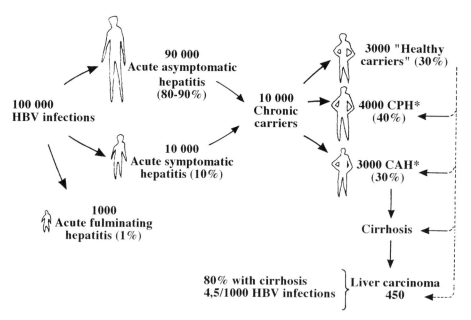

FIGURE 97

Evolution of B hepatitis infections. (Adapted from Vasseur, M., *Les Virus Oncogènes,* Hermann Éditeurs, Paris, 1989, 232. With permission.)

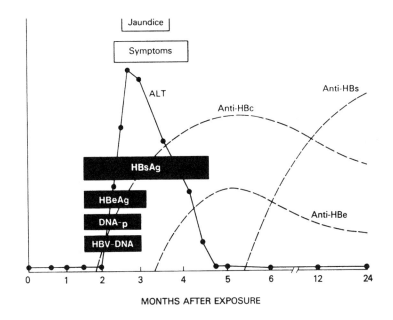

FIGURE 98

The clinical, serum biochemical, and serologic course of a typical acute hepatitis B. HBsAg, hepatitis surface antigen, HBeAg; hepatitis B e antigen; DNA-p, DNA polymerase activity; HBV, hepatitis B virus; ALT, alanine aminotransferase; Anti-HBc, antibody to hepatitis core antigen; Anti-HBs, antibody to HBsAg (From Hoofnagle, J.H. and Di Bisceglie, A.M., in *Antiviral Agents of Viral Diseases of Man*, 3rd ed., Galasso, G.J., Whitley, R.J., and Merigan, T.C., Eds., Raven Press, New York, 1990, 145-459. © Lippincott Williams & Wilkins. With permission.)

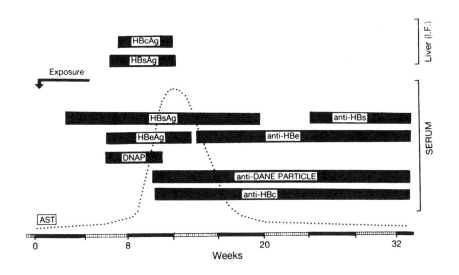

FIGURE 99

Virological events in acute hepatitis B in relation to serum aminotransferase (AST) peak. (From Eddleston, A.L.W.F., in *Oxford Textbook of Medicine,* 2nd ed. Vol. 1, Weatherall, D.J., Ledingham, L.G.G., and Warrell, D.A., Eds., Oxford University Press, Oxford, 1987, 12.214. By permission of Oxford University Press.)

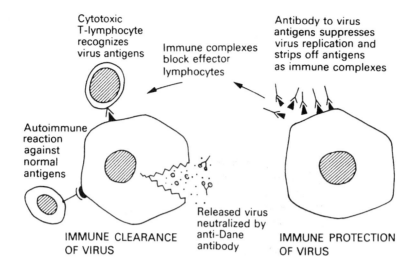

FIGURE 100

Immune mechanisms acting to kill virus-infected cells and eliminate the infection *(left)* and other immune mechanisms which may interfere with this process *(right).* (From Eddleston, A.L.W.F., in *Oxford Textbook of Medicine,* 2nd ed. Vol. 1, Weatherall, D.J., Ledingham, L.G.G., and Warrell, D.A., Eds., Oxford University Press, Oxford, 1987, 12.216. By permission of Oxford University Press.)

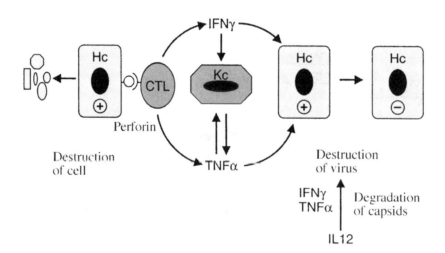

FIGURE 101

Role of cytokines in virus elimination. The cytotoxic T cell immune response induces lysis of infected cells and has also an antiviral effect leading to virus destruction without lysis of hepatocytes. CTL, cytotoxic T lymphocyte; Hc, hepatocyte; Kc, Kupffer cell. (Adapted from Zoulim, F. and Trepo, C., *Virologie,* 1, 197, 1997. With permission of John Libbey Eurotext.)

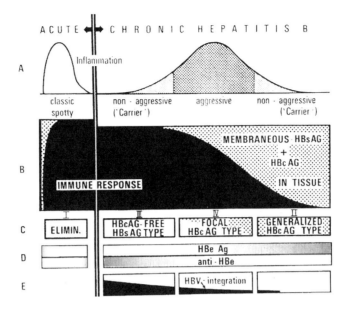

FIGURE 102

HBV infection: Correlation between (A) inflammatory reaction in liver tissue, (B) inverse reaction of immune responsiveness and display of BHcAg plus membraneous HBsAg, (C) patterns of viral antigen expression, (D) HBeAg/anti-HBe system, and (E) integration of HBV DNA. (From Bianchi, L. and Gudat, F., in *Chronic Hepatitis,* Liaw, Y.-F., Ed., Excerpta Medica, Amsterdam, 1985, 45. With permission by L. Bianchi.)

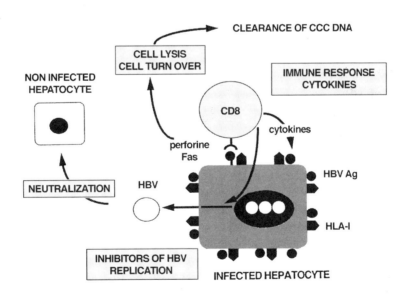

FIGURE 103

Pathogenesis of chronic hepatitis B. HBV is not cytopathogenic. Viral replication leads to the production of viral particles that continuously infect other cells and to the expression of viral antigens at the cell membrane. These antigens are presented by MHC class I molecules to cytotoxic T cells that kill infected cells via the fas/perforin pathway. Cytotoxic T cells may also exhibit a direct antiviral effect by the secretion of antiviral cytokines (TFN alpha, IFN gamma, IL12) independently of cell lysis. It was also suggested that viral clearance and curing of infected hepatocytes may result from hepatocyte division. Chronic hepatitis B results from the persistence of viral replication and the continuing attack of infected hepatocytes which are not sufficient to clear all infected cells. (From Zoulim, F., *Antiviral Res.,* 44, 1, 1999.)

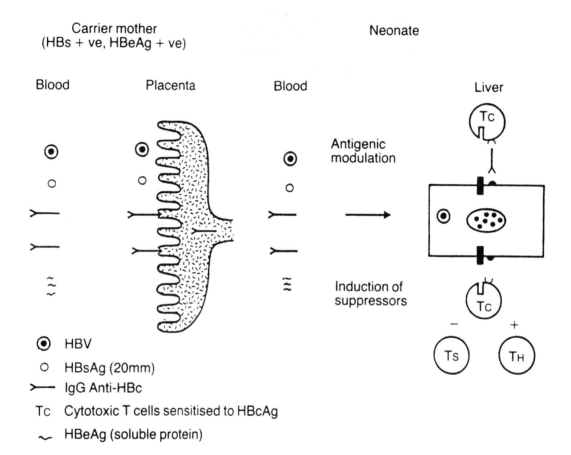

FIGURE 104

Neonatal HBV infection: postulated mechanism of viral persistence. It is proposed that HBe antigen crosses the placenta from the maternal blood and, in neonates, induces tolerance to HBe antigen which is one of the targets of the immune response. Maternal IgG anti-HBc also crosses the placenta into the foetal circulation. HBV infection of the neonatal liver is thus facilitated as maternal IgG blocks recognition of virus-infected cells by cytotoxic T cells. Early exposure to soluble virus protein (HBe) may induce a state of antigenic tolerance to the virus with specific suppressor cells inhibiting the host defense mechanism. (From Thomas, H.C., in *New Antiviral Strategies,* Norrby, S.R., Mills, J., Norrby, E., and Whitton, L.J., Eds., Churchill Livingstone, Edinburgh, 1988, 52. With permission.)

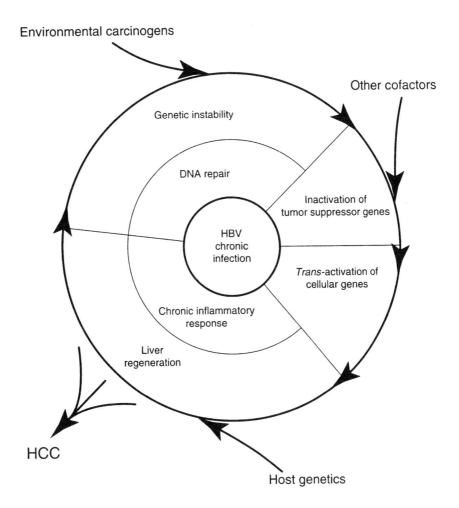

FIGURE 105

Model for the role of hepatitis B virus (HBV) in development of liver cancer. Chronic infection by HBV is central to the process. The resulting inflammatory response and liver regeneration introduce the potential of errors during DNA replication. The HBV X (HBx) protein interacts with X-associated protein 1 (XAP-1), a cellular DNA repair protein. If viral replication were to introduce the ability of the cell to repair damaged DNA, mutations would accumulate over time, increasing the likelihood that tumor suppressor genes such as *p53* would get functionally inactivated. HBx can *trans*-activate cellular genes, perhaps through a transcription-coupled repair process; affected cell genes might contribute to carcinogenesis. Environmental carcinogens are important cofactor in certain areas of the world and would be especially potent if viral infection had crippled the cellular DNA repair system. Host genetics influences the host response to HBV infection. When the appropriate genetic changes accumulate over years, hepatocellular carcinoma (HCC) occurs. (Reprinted from *Trends Microbiol.*, 4, 119-124, 1996. Butel, J.S., Lee, T.-H., and Slagle, B.L., Is the DNA repair system involved in hepatitis-B-virus-mediated hepatocellular carcinogenesis? © 1996, with permission of Elsevier Science.)

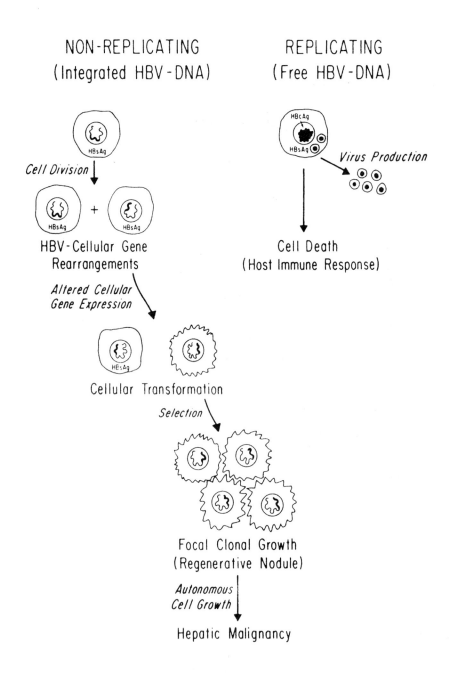

FIGURE 106

Possible events following HBV infection and integration of HBV DNA into the liver cell genome
that might ultimately lead to the development of HCC. On the right are shown HBV carriers who
continue to replicate the virus and demonstrate continued liver disease; on the left are shown
HBV carriers who do not replicate the virus and show little or no liver disease. (From Shafritz,
D.A. and Hadziyannis, S.J., in *Advances in Hepatitis Research*, Chisari, F.V., Ed., Masson
Publishing USA, New York, 1984, 241. © Masson Editeur. With permission.)

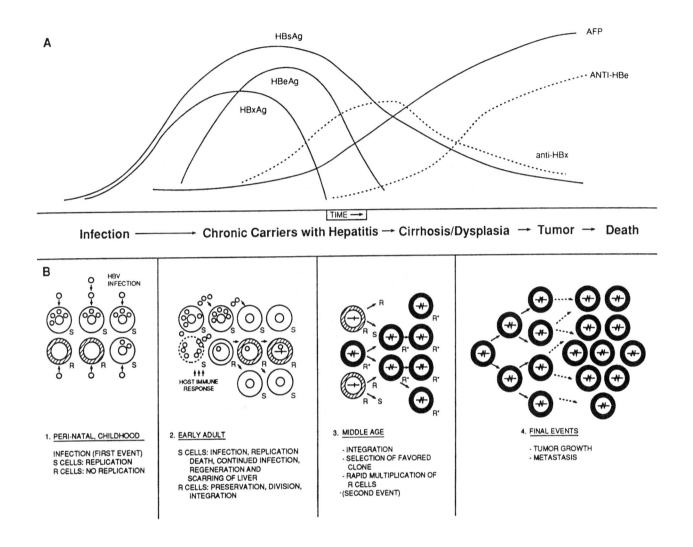

FIGURE 107

Frequent serological profiles of HBV chronic carriers. Such carriers often have detectable HBsAg in serum for years or decades after the initial infection. Many patients with lengthy periods of virus replication also have detectable levels of HBeAg and/or HBxAg in blood. During chronic infection, many individuals lose HBxAg and seroconvert to anti-HBx while clearing virus from serum. This seroconversion is often followed by loss of HBeAg and development of anti-HBeAg. Some carriers experience one or more episodes of chronic hepatitis, which can develop into cirrhosis and dysplasia. These patients are at high risk for the development of hepatocellular carcinoma, which is often accompanied by rises in serum alpha-feto-protein (AFP). (B) Model for PHC associated with chronic HBV infection. A susceptible liver is rich in hepatocytes that support HBV replication (S cells). During chronic infection accompanied by bouts of hepatitis, S cells are recognized and lysed by immunological mechanisms, and the liver becomes increasingly rich in R cells, which do not support virus replication. Such R cells have integrated HBV DNA and may in time be selected for expansion. (From Blumberg, B.S., Millman, I., Venkateswaran, P.S., and Thyagarajan, S.P., *Cancer Detect. Prev.,* 14, 195, 1989. Reprinted by permission of Blackwell Science, Inc.)

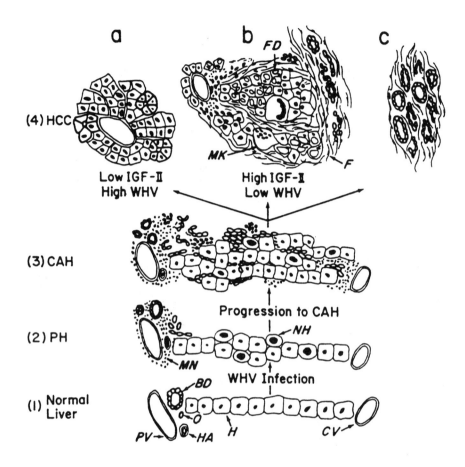

FIGURE 108

Proposed progressive stages of woodchuck hepatitis infection leading to generation of biochemically and histologically heterogeneous primary hepatocellular carcinomas in woodchucks. BD, bile duct; CAH, chronic active hepatitis; CV, central vein; F, fibroblast; FD, fatty degeneration; H, hepatocyte; HA, hepatic artery; IGF-II, insulin-like growth factor II; MK, megakaryocyte; MN, mononuclear cell; NH, necrotic hepatocyte; O, oval cell; PV, portal vein; WHV, woodchuck hepatitis virus.
Line 1. Normal liver architecure comprising portal tracks *(left)* connected by single cell-plates of hepatocytes and central hepatic veins *(right)*. Line 2. After establishment of WHV persistent hepatitis, portal hepatitis develops. Line 3. Portal hepatitis progresses to chronic active hepatitis. Line 4. Sustained by the dynamics of protracted infection, inflammation, cell necrosis, and growth factor production, heterogeneous types of tumors arise. Illustrated are (a) low-IGF-II, high-WHV tumors, (b) high-IGF-II, low WHV tumor with more dysplastic features, or very rare instances of cholangiocarcinoma. (From Fu, X.-X., Su, C.Y., Lee, Y., Hintz, R., Biempica, L. Snyder, R., and Rogler, C.E., *J. Virol., 62*, 3422, 1988. With permission.)

A. GENUS *DELTAVIRUS*

Circular ssRNA
Enveloped, no capsid

Hepatitis delta virus (HDV) is the only member of a nonassigned genus. The virus is defective, needs the presence of coinfecting HBV, and borrows its envelope from it. Particles are spherical, 34 nm in diameter, and consist of an envelope of HBsAg and capsid-less helical nucleoprotein with circular ssRNA. The genome resembles that of plant viroids and may be considered as satellite RNA with an envelope. Delta hepatitis seems to be prevalent in the Mediterranean area and resembles B hepatitis in transmission (by blood and blood products) and tendency toward chronicity. Coinfection by HBV and delta virus results in acute B and D hepatitis; superinfection of B hepatitis with HDV results in severe disease with a tendency toward both fulminant hepatitis and chronic disease.

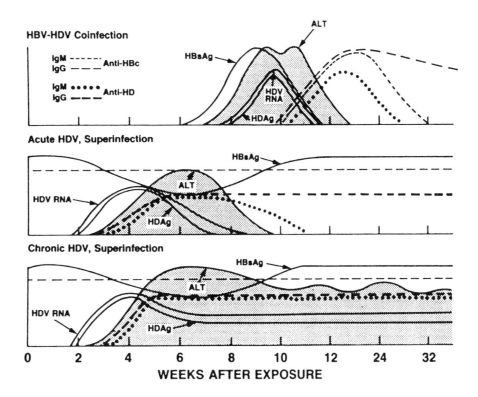

FIGURE 109

Clinical and serological events in typical cases of type D hepatitis resulting from acute HBV and HDV infection *(top),* acute HDV superinfection of an HBsAg carrier *(middle)* and HDV superinfection progressing to chronic type D hepatitis in an HBsAg carrier *(bottom).* (From Purcell, R.H., Hoofnagle, J.H., Ticehurst, J., and Gerin, J.L., in *Diagnostic Procedures for Viral, Rickettsial, and Chlamydial Infections,* 6th ed., Schmidt, N.J. and Emmons, R.W., Eds., American Public Health Association, Washington, 1989, 957. © APHA, with permission.)

6.IX. HERPESVIRIDAE

Linear dsDNA
Cubic, enveloped

Members of this large family occur in all branches of vertebrates and are classified into 10 genera and three subfamilies *(Alpha-, Beta-, Gammaherpesvirinae)* according to DNA content and structure. Particles are enveloped, 120-200 nm wide, and contain an icosahedral capsid 100-110 nm in diameter. Herpesviruses are highly adapted to their hosts and host ranges are generally narrow. They include many pathogens, tend to cause latent infections in particular cell types, and have evolved a state of stable, persistent parasitism in many mammals. Except for very young or immunocompromised hosts, infections are rarely lethal.

A. *ALPHAHERPESVIRINAE*

This subfamily includes the genera *Simplexvirus* and *Varicellovirus.* Both comprise numerous viruses which resemble each other closely in mechanisms of pathogenesis and are able to establish latent infections in sensory ganglia. Among the many diseases caused by these viruses, only herpes simplex, varicella-zoster, Aujeszky's disease, and Marek's disease are illustrated. Although Fig. 112 is meant to illustrate herpes simplex virus infections only, it also applies to varicella-zoster virus infections.

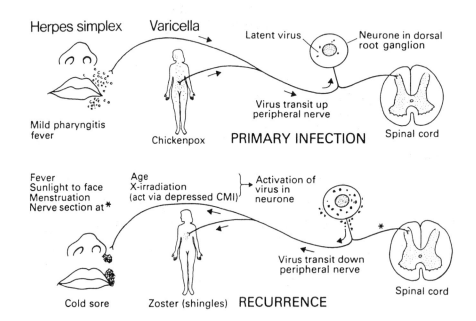

FIGURE 110

Mechanisms of herpes simplex and varicella-zoster latency and reactivation in man. (From Mims, C.A. and White, D.O., *Viral Pathogenesis and Immunology,* Blackwell Scientific Publications, Oxford, 208, 1984. With permission.)

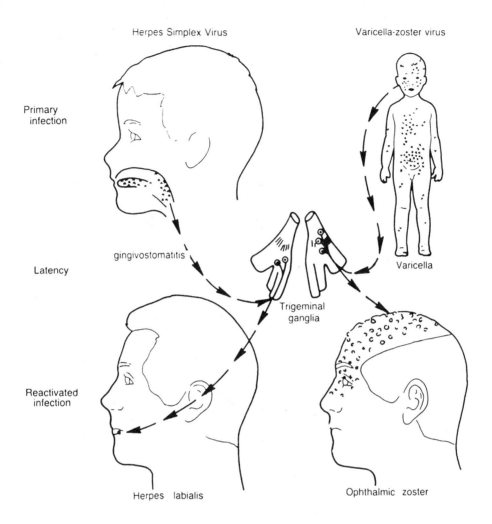

Herpes Simplex Virus

Varicella-zoster virus

Primary infection

gingivostomatitis

Latency

Trigeminal ganglia

Varicella

Reactivated infection

Herpes labialis

Ophthalmic zoster

Neural latency of herpes viruses

FIGURE 111

Diagram of primary childhood infection, trigeminal ganglia latency, and after-life inactivation of herpes simplex and varicella-zoster viruses. Primary infection with herpes simplex is most often a gingivostomatitis or pharyngitis during which virus ascends along sensory nerves and becomes latent in the third division of the trigeminal ganglia. Later activation can cause lesions in the oropharynx or the mucocutaneous junction. Primary infection of varicella gives widespread lesions over the trunk and face, and latency presumably can develop in sensory ganglia. The first division of the trigeminal ganglia is the most frequently affected, and later activation gives ophthalmic zoster, with characteristic dermatomal distribution of lesions over the first division of the trigeminal nerve. (From Johnson, R.T., *Viral Infections of The Central Nervous System,* Raven Press, New York, 1982, 134. © Lippincott Williams & Wilkins. With permission.)

HERPES SIMPLEX

Acute infection by herpes simplex virus (HSV) is characterized by vesicular eruptions on the skin or mucosa followed by latency. Type 1 causes the universally known herpes labialis and gingivostomatitis, but also keratoconjunctivitis, meningoencephalitis, and disseminated visceral forms. Type 2 mostly causes genital infections. After infection, the virus travels via axons toward sensory ganglia where it persists as an episome. Type 1 has a preference for the trigeminal ganglion; type 2 tends to establish itself in sacral ganglia. With or without provocation (e.g., sunlight, fever), the virus is reactivated, spreads down nerve fibers to its portal of entry and causes there recurrent infections. HSV 2 causes extremely serious infections in the fetus and newborns and is a possible cofactor in the development of cervical cancer.

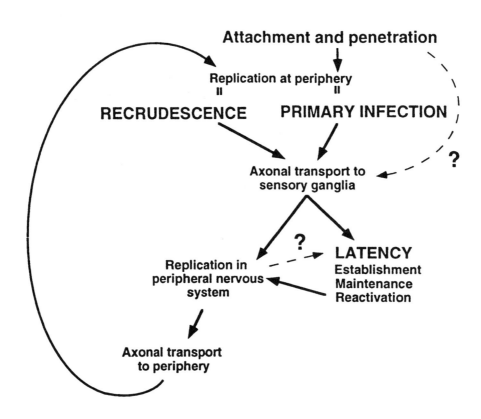

FIGURE 112

Herpes simplex virus (HSV) infection in animals. After attachment and penetration, replication takes place in epithelial cells at the periphery. Virus then infects nerve cell termini and is delivered by fast axonal transport to the sensory ganglia where it can either replicate or establish latency. Virus can be reactivated by certain stimuli which stress the neuronal cells. Reactivation may lead to replication in the peripheral nervous system, following which virus may be transported either back to the periphery or to the central nervous system. Replication at the periphery following reactivation is called recrudescence, and may once again lead to axonal transport to the sensory ganglia. Immune mediated clearance of replicating virus occurs at the periphery and in the peripheral nervous system. Solid arrows denote pathways. Dashed arrows with question marks denote possible alternative pathways. (From Fraser, N.W. and Valya-Nagy, T., *Microbial Pathogen.*, 15, 83, 1993. With permission.)

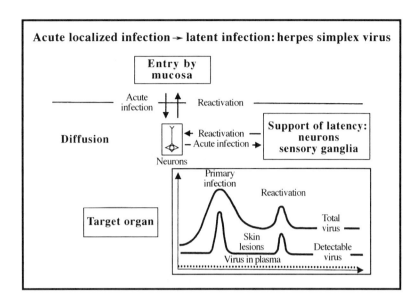

FIGURE 113

Herpes simplex: an example of a local infection evolving toward latency. The virus replicates in epithelial cells and then migrates toward its site of latency using nerve cells of sensory ganglia. The virus persists in noninfectious form until reactivation and is constantly absent from the plasma compartment. (Adapted from Maréchal, V., Dehée, A., and Nicolas, J.-C., *Virologie,* 1 (special issue), 11, 1997. With permission of John Libbey Eurotext.)

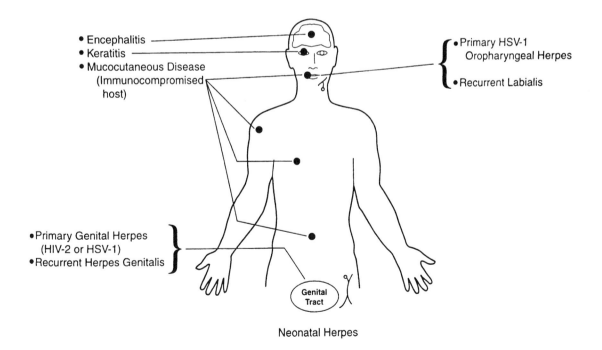

FIGURE 114

Spectrum of HSV-caused disease. (From Whitley, R.J., in *Fields Virology,* 3rd ed., Vol. 2, Fields, B.N., Knipe, D.M., and Howley, P.M., Eds.-in-chief, Lippincott-Raven, Philadelphia, 1996, 2297-2342. © Lippincott Williams & Wilkins. With permission.)

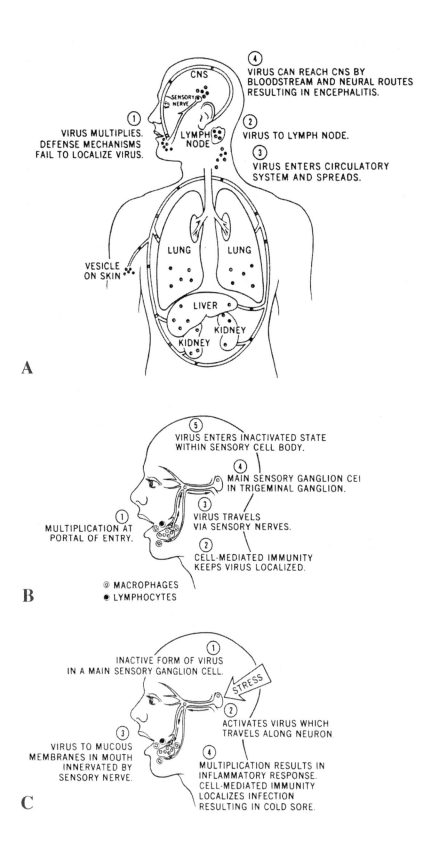

FIGURE 115

(A) Systemic disease due to herpes simplex virus. (B) Steps leading to latency. (C) Activation of recurrent disease. (Reprinted with permission from Rapp, F., in *Virology and Rickettsiology*, Vol. I, Part 2, Hsiung, G.-D. and Green, R.H., Eds., CRC Press, Boca Raton, 1978, 75. Copyright CRC Press, Boca Raton, Florida.)

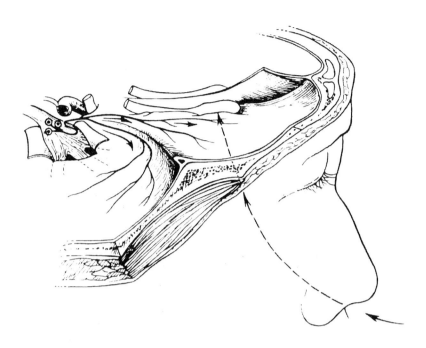

FIGURE 116

Possible anatomical explanations for the orbital-frontal and temporal localization of herpes simplex virus encephalitis. Direct invasion of the olfactory bulb *(right)* could produce orbital-frontal infection, with spread to the adjacent temporal lobes. Small sensory fibers from the trigeminal ganglia *(left)* send fibers to the basilar dura of the anterior and middle fossae. (From Johnson, R.T., *Viral Infections of the Central Nervous System,* New York, 1982, 137. © Lippincott Williams & Wilkins. With permission.)

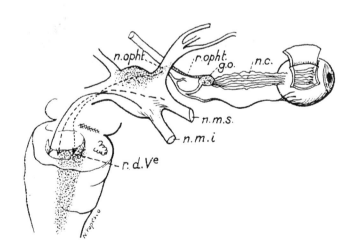

FIGURE 117

Lesions resulting from inoculation of herpes simplex virus on the rabbit cornea. Lesions are illustrated in the ciliary ganglion (g.o.), the ophthalmic division of the trigeminal nerve (n. opht.) and within the descending trigeminal tract of the brainstem (r.d.V^e). The latter lesions occupy the ventral portion of the tract, which contains fibers of the ophthalmic division. (Historical drawing by B. N. Popesco, from Marinesco, G., *Rev. Neurol.,* 1, 1, 1932. © Masson Editeur. With permission.)

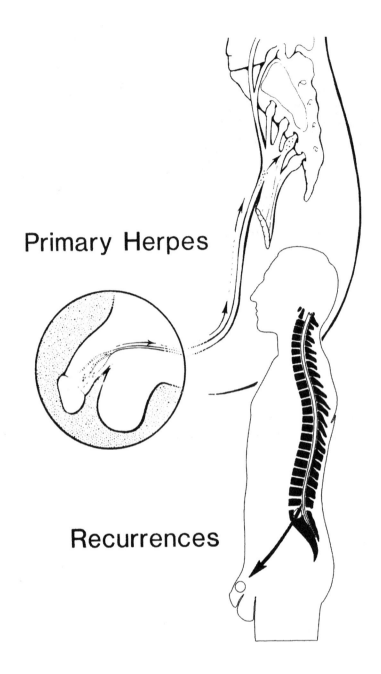

Primary Herpes

Recurrences

FIGURE 118

Migration of herpes simplex virions from genital lesions to sacral ganglia during primary herpes and from the ganglia back to the skin during recurrences. (Modified from Balfour, H.H. and Heussner, R.C., *Herpes Diseases and Your Health,* University of Minnesota Press, Minneapolis, 1984, 38. With permission by H.H. Balfour.)

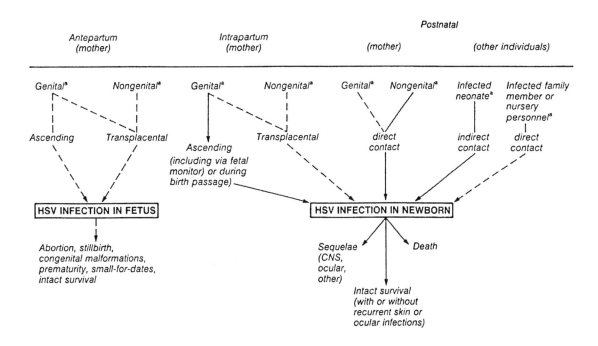

FIGURE 119

Possible sources and modes of transmission of herpes simplex infection in the fetus and newborn. [a]Clinically manifest of subclinical infection in mother, the infection may have been first acquired prior to conception. Continuous line, substantial evidence; dashed line, poorly substantiated. (From Nahmias, A.J., Keyserling, H.L., and Kerrick, G.M., in *Infectious Diseases of The Fetus and The Newborn Infant,* 2nd ed., Remington, J.S. and Klein, J.O., Eds., W.B. Saunders, Philadelphia, 1983, 636. With permission.)

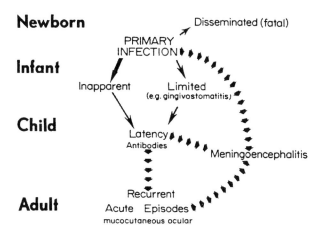

FIGURE 120

Patterns of infection with herpes simplex virus. Primary exposure occurs during infancy and may produce fatal disseminated infection, limited overt disease, or inapparent infection, followed by latent infection of sensory ganglia. During adult life, reactivation may occur, with limited mucocutaneous or ocular involvement or spread to the nervous system. (Reprinted with permission from Monath, P., in *Virology and Rickettsiology,* Vol. I, Part 2, Hsiung, G.-D. and Green, R.H., Eds., CRC Press, Boca Raton, 1978, 261. Copyright CRC Press, Boca Raton, Florida.)

A. Peripheral infection

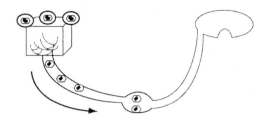

B. Acute ganglion infection

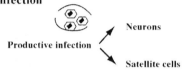

Productive infection → Neurons

→ Satellite cells

C. Latency

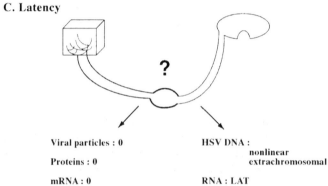

?

Viral particles : 0   HSV DNA :

nonlinear
extrachromosomal

Proteins : 0

mRNA : 0   RNA : LAT

D. Reactivation

Stress, fever, hormonal factors, UV, trauma

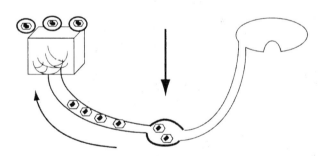

FIGURE 121

Herpes simplex infection *in vivo.* LAT, latency-associated transcripts. (Adapted from Rozenberg, F., in *Virologie Moléculaire Médicale,* Seigneurin, J.-M. and Morand, P., Eds., Tec & Doc Lavoisier, Paris, and EM Inter, Cachan, 1997, 143. With permission.)

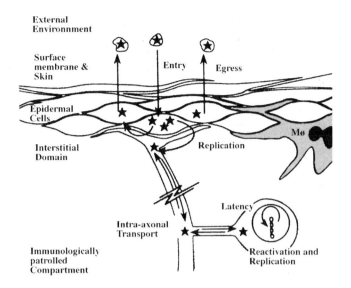

FIGURE 122

The HSV survival machine. The diagram illustrates the anatomical sites infected by HSV and indicates the immunoprivileged nature of HSV infection. Immunologically patrolled areas are indicated. Sites where virus evades immune recognition are: (a) the axon, when the virus is travelling intra-axonally; (b) the neuron, during virus latency; and (c) the epidermis, following the initial stages of virus recurrences. Mϕ indicates the macrophage. The immune system eventually controls the infection in the epidermis but not before egress of virus to infect another individual has occurred. (From Nash, A.A. and Cambaropoulos, P., Sem. Virol., 4, 181, 1993. With permission.)

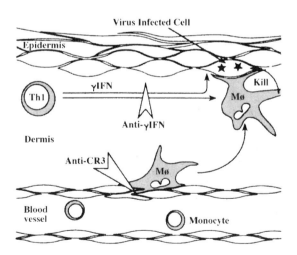

FIGURE 123

Role of CD4 cells in recruiting and arming macrophages to mediate anti-HSV immunity. Anti-CR3 blocks emigration of monocytes into the skin, and anti-IFN-ϕ blocks the activation of macrophages (Mγ). Both these events markedly reduce clearance of virus from the epidermis (Cambaropoulos, P., Nash, A.A., unpublished observations). (From Nash, A.A. and Cambaropoulos, P., Sem. Virol., 4, 181, 1993. With permission.)

VARICELLA-ZOSTER

Varicella, often referred to as chickenpox, is a highly contagious human childhood disease. Contrary to other herpesviruses, varicella-zona virus (VZV) is transmitted by respiratory droplets. A prodromal stage with fever is followed by a maculopapular rash which first appears on the trunk. Lesions progress to vesicles, pustules, and scabs. The disease is benign except in immunocompromised individuals. The virus then becomes latent in dorsal root and trigeminal sensory ganglia. Herpes zoster (Greek, zoster, "girdle") is the recurrent form of varicella. After decades, the virus may be reactivated and migrates then within a particular sensory nerve to the skin where it causes painful blisters in the corresponding skin segment (dermatome). T lymphocyte-mediated immunity is critical for preserving the balance between host and latent VZV. Herpes zoster is particularly common and severe in immunocompromised individuals.

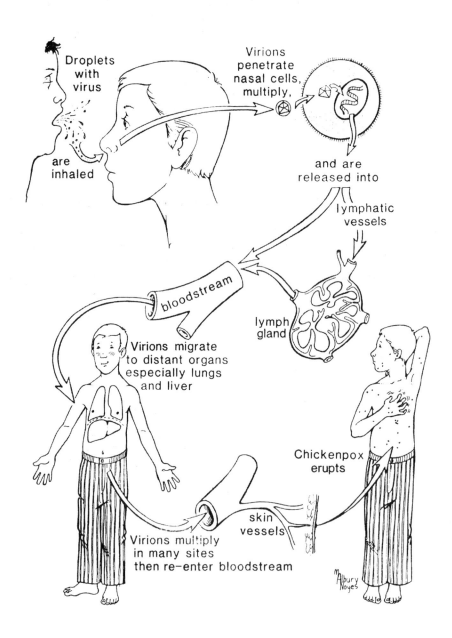

FIGURE 124

Path of the chickenpox virus. (From Balfour, H.H. and Heussner, R.C., Herpes Diseases and Your Health, University of Minnesota Press, Minneapolis, 1984, 79. With permission by H.H. Balfour.)

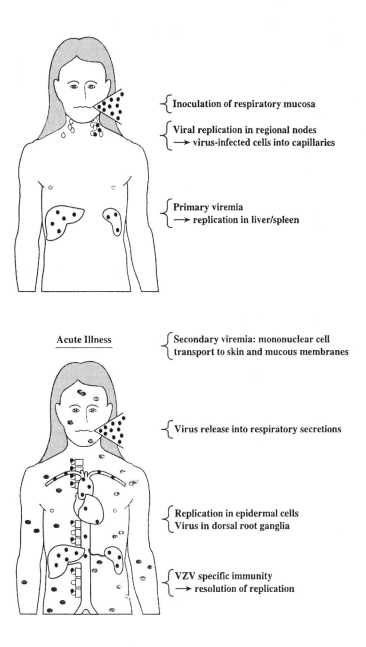

FIGURE 125

Pathogenesis of primary infection with varicella-zoster virus. Primary infection with VZV starts in the respiratory mucosa. During the incubation period, which lasts from 10 to 21 days, the virus probably spreads to regional lymph nodes and then causes a primary viremia with associated viral replication in liver and spleen. A second cell-associated viremic phase begins at 24 to 96 hr and results in the transport of the virus to the skin and respiratory mucosal sites. Replication in epidermal cells causes the characteristic rash of varicella. The induction of VZV-specific immunity is required to terminate viral replication. The virus gains access to cells of the trigeminal and dorsal root ganglia during primary infection and established latency. (From Arvin, A.M., in Fields Virology, 3rd ed., Vol. 2, Fields, B.N., Knipe, D.M., and Howley, P.M., Eds-in-chief., Lippincott-Raven, Philadelphia, 1996, 2547-2587. © Lippincott Williams & Wilkins. With permission.)

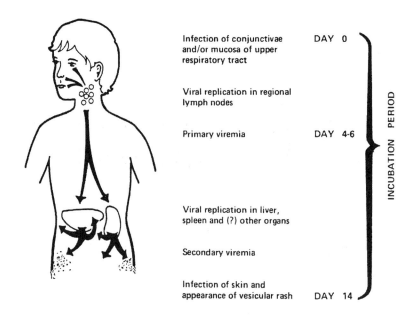

FIGURE 126

Pathogenesis of wild-type varicella in children. The incubation period comprises a primary and a secondary viremia. The schema illustrates why varicella-zoster immune globulin must be administered within 3 to 4 days of infection in order to be effective, that is, while it can abrogate the primary viremia. The typical exanthem of varicella appears a few days after the second viremia, as the virus travels from the capillaries to the epidermis. The infected human becomes contagious near the end of the incubation period before the onset of rash. (From Grose, C., Pediatrics, 68, 735, 1981.)

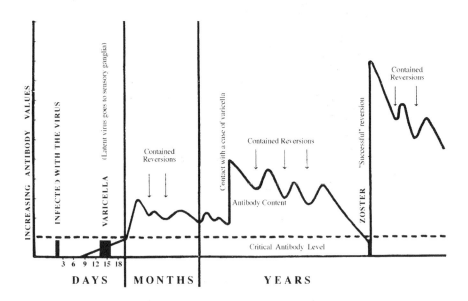

FIGURE 127

Model of zoster pathogenesis. After primary infection and establishment of latency, VZV reactivates numerous times. Most reactivations are contained by the host, perhaps by an immune-mediated mechanism. Once the level of host resistance drops below a critical threshold, reactivation of the virus leads to zoster. Following the zoster episode, host defenses can once again contain periodic recurrent infections. (Adapted from Hope-Simpson, R.E., Proc. Roy. Soc. Med., 58, 9, 1965. With permission.)

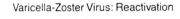

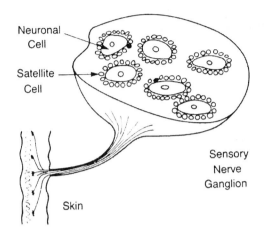
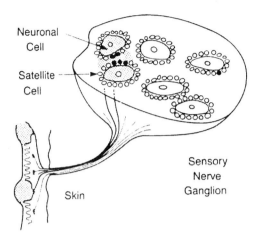

FIGURE 128

Persistence of latent VZV within a sensory ganglion and hypothetical pattern of infection during reactivation. Large cells are neurons; small cells are nonneuronal (including satellite, endothelial, and fibroblast-like cells). Blackened cells contain latent virus. Productive virus replication is implied by stippling. Reactivated viruses spread across the ganglia to multiple satellite cells and neurons and subsequently to the corresponding dermatome. (Lower half of Fig. 3 from Straus, S.E., J. Am. Med. Assoc., 262, 3455-3458, 1989. With permission.)

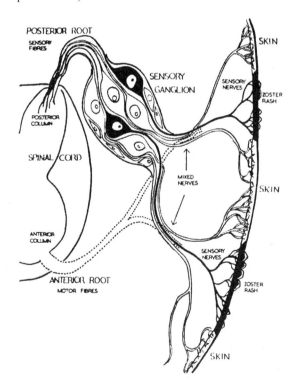

FIGURE 129

Illustration of the pathogenesis of zoster. Affected neurons and sensory nerves are in black. (From Hope-Simpson, R.E., Proc. Roy. Soc. Med., 58, 9, 1965. With permission.)

AUJESZKY'S DISEASE

Also called "pseudorabies", the disease is a rapidly and highly lethal encephalitis of cattle and newborn pigs. Adult pigs are the reservoir of the virus. After respiratory infection, the virus replicates in nasopharynx, tonsils, and lungs. Adult pigs usually develop fever and pneumonia. The virus reaches the central nervous system via the neural route (trigeminal, glossopharyngeal, or olfactory nerves). Latency is achieved by persistence within sensory (trigeminal) ganglia.

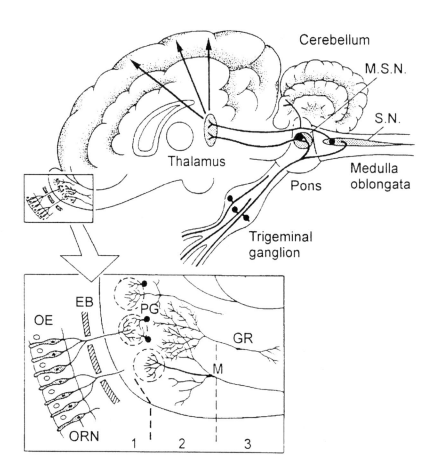

FIGURE 130

Synaptically linked sensory neurons of the olfactory and trigeminal routes by which the virus can spread transneuronally. The enlargement shows details of the olfactory route. Bipolar olfactory receptor neurons (ORN) in the olfactory epithelium (OE) are the first-order neurons of the olfactory route. They send axons through the ethmoid bone (EB) into the olfactory bulb, where they end in spherical neurophils called glomeruli (dashed circles). Here the axons synapse on second-order neurons, i.e., periglomerular neurons (PG) and mitral neurons (M). Pseudo-unipolar neurons in the trigeminal ganglion are the first-order neurons in the trigeminal route. They receive afferent axons from the ophthalmic nerve, the maxillary nerve, and the mandibular nerve. These trigeminal neurons send efferent axons through the trigeminal nerve that end in the brain stem and medulla oblongata. The majority of these axons synapse on second-order neurons of the main sensory nucleus (M.S.N.) of the trigeminal nerve located in the brain stem at the pons cerebri. Another part of these axons end in the spinal nucleus (S.N.) in the medulla oblongata. (From Mulder, W., Pol, J., Kimman, T., Kok, G., Priem, J., and Peeters, B., J. Virol., 70, 2191, 1996. With permission.)

MAREK'S DISEASE

Its agent, which has only recently been classified as an alphaherpesvirus, causes a polymorphic infection in fowl characterized by mononuclear cell infiltrations and lymphomas in nerves, visceral organs, muscle, and skin. The disease is a major cause of mortality in chicken farms. Up to 70% of chickens die within 6 to 8 weeks and few, if any, commercial poultry farms escape infection. Infection is usually by inhalation of infected feather follicles. The virus multiplies in lymphoid tissues and is found in thymus, spleen, and the bursa of Fabricius, causing a massive lymphocytic invasion of internal organs and nerve trunks. The virus replicates in, and is shed from, the feather follicle epithelium. Some isolates are not oncogenic. Virus genomes can persist in cells in integrated and nonintegrated form.

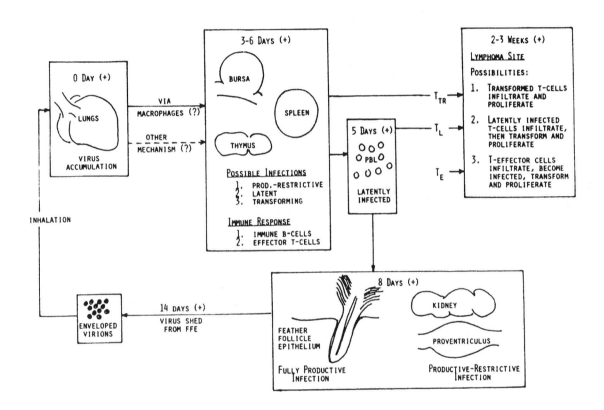

FIGURE 131

Sequential events in the pathogenesis of Marek's disease. Fowl are usually infected by inhalation. Viruses replicate in lymphoid tissues (bursa of Fabricius, thymus, spleen). PBL, peripheral blood lymphocytes. Lymphomas and inflammatory reactions develop after 2-3 weeks. T_{TR}, T_L, T_E refer to proposed subgroups of T lymphocytes which are, respectively, the targets of transformation, latent infection, or are effector cells in cell-mediated immunity responses. (Reprinted with permission from Calnek, B.W., in Oncogenic Herpesviruses, Vol. I, Rapp, F., Ed., CRC Press, Boca Raton, 1980, 103. Copyright CRC Press, Boca Raton, Florida.)

B. *BETAHERPESVIRINAE*

Cytomegaloviruses are the prototype members of this subfamily. They occur in most mammals and produce a particular cytopathic effect, characterized by the presence of large cells with target-like nuclei distended by huge inclusion bodies. Macrophages and salivary gland cells have been incriminated as the sites of latency. Human cytomegalovirus is a common parasite of humans. Infection is congenital or perinatal, by sexual contact, or iatrogenic. In healthy adults, the disease is usually asymptomatic because the virus is under strict immunological control by NK cells or T lymphocytes, especially of the CD8+ subset. The virus persists throughout life by evading antigen presentation by the major histocompatibility complex (MHC) class I. Congenital infections range from the asymptomatic process to jaundice, purpura, anemia, and death. Immunodeficient adults, for example AIDS patients, develop a similarly wide range of symptoms.

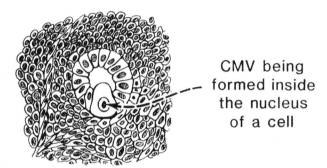

FIGURE 132

A microscopic section from a salivary gland infected with CMV. The virus has invaded a central duct, causing the cell to enlarge and forming a mass within the nucleus that consists of many incomplete viral particles. (From Balfour, H.H. and Heussner, R.C., *Herpes Diseases and Your Health*, University of Minnesota Press, Minneapolis, 1984, 126. With permission by H.H. Balfour.)

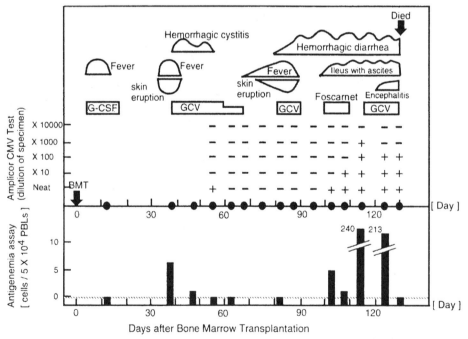

FIGURE 133

Clinical course for an 18-year-old female patient. The results of the antigenemia were expressed as numbers of positive cells per 5 x 104 peripheral blood leukocytes (PBLs). CMV was treated with intravenous ganciclovir (GCV) at 5 mg/kg of body weight twice a day and intravenous foscarnet at 200 mg/kg/day. G-CSF, granulocyte colony-stimulating factor; +. positive; -, negative. (From Hiyoshi, M., Tagawa, S., Takubo, T., Tanaka, K., Nakao, T., Higeno, Y., Tamura, K., Shimaoka, M., Fujii, A., Higashikata, M., Yasui, Y., Kim, T., Hiraoka, A., and Tatsumi, N., J. Clin. Microbiol., 35, 2692, 1997. With permission.)

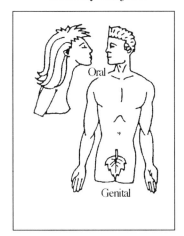
Healthy subjects

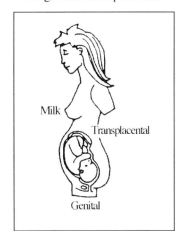
Congenital and perinatal

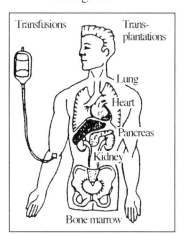
Iatrogenic

A

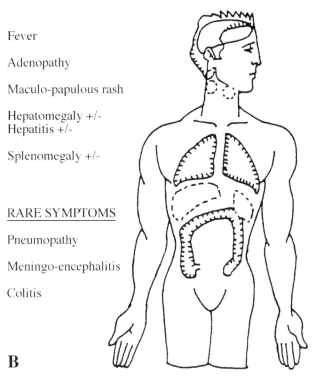

Fever

Adenopathy

Maculo-papulous rash

Hepatomegaly +/-
Hepatitis +/-

Splenomegaly +/-

RARE SYMPTOMS

Pneumopathy

Meningo-encephalitis

Colitis

B

SAMPLES/EXAMS

Seroconversion or
elevation of antibodies
  - CF
  - ELISA (IgM)

SEVERE FORMS

Viremia
Antigenemia
PCR (leukocytes and/or plasma)
BAW or lung biopsy
Colon biopsy
Liver biopsy

  - virus isolation
  - immunocytochemistry
  - PCR-CMV

FIGURE 134

(A) Transmission of cytomegalovirus. (B) Symptoms of CMV infection in adults and required tests. CF, complement fixation; BAW, broncho-alveolar washings. (Adapted from Colimon, R. and Minjolle, S., in Virologie Moléculaire Médicale, Seigneurin, J.-M. and Morand, P, Eds., Tec & Doc Lavoisier, Paris, and EM Inter, Cachan, 1997, 169. With permission.)

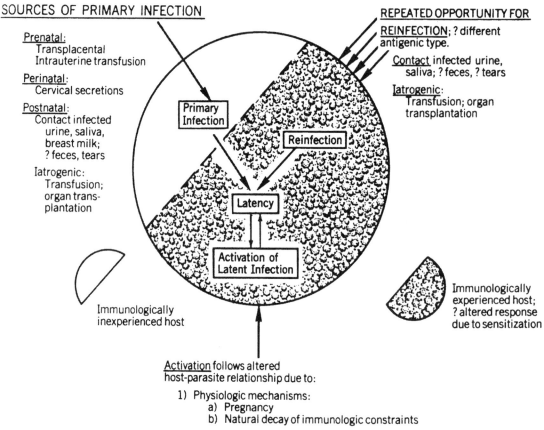

SOURCES OF PRIMARY INFECTION

<u>Prenatal</u>:
　Transplacental
　Intrauterine transfusion

<u>Perinatal</u>:
　Cervical secretions

<u>Postnatal</u>:
　Contact infected
　urine, saliva,
　breast milk;
　? feces, tears

Iatrogenic:
　Transfusion;
　organ trans-
　plantation

<u>REPEATED OPPORTUNITY FOR</u>
<u>REINFECTION</u>; ? different
antigenic type.

<u>Contact</u> infected urine,
　saliva; ? feces, ? tears

<u>Iatrogenic</u>:
　Transfusion; organ
　transplantation

Primary
Infection

Reinfection

Latency

Activation of
Latent Infection

Immunologically
inexperienced host

Immunologically
experienced host;
? altered response
due to sensitization

<u>Activation</u> follows altered
host-parasite relationship due to:

　1) Physiologic mechanisms:
　　　a) Pregnancy
　　　b) Natural decay of immunologic constraints

　2) Concurrent debilitating disease

　3) Iatrogenic: immunosuppressive drugs
　　　or operative procedures

FIGURE 135

Natural history of human CMV infection. (From Weller, T.H., New England J. Med., 285, 267, 1971. © 1971, Massachusetts Medical Society. All rights reserved.)

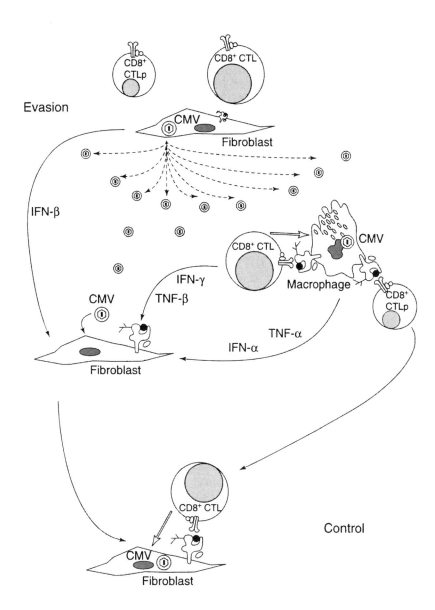

FIGURE 136

Host factors that counteract cytomegalovirus (CMV) evasion from major histocompatibility complex (MHC) class I presentation. CMV genes prevent antigen presentation during permissive infection of fibroblasts (the efficacy of MHC class I presentation is represented by the size of MHC class I molecules shown on the cell surface). Unfilled arrows indicate presentation resulting in lysis of CMV-infected cells by CD8$^+$ lymphocytes (CTLs). Interferon β (IFN- β) secreted from infected fibroblasts, CD8$^+$ T-cell-derived cytokines [IFN-γ and tumor necrosis factor β(TNF- β)] and cytokines released from CMV-infected macrophages (IFN- α, TNF- α) restore antigen presentation to CD8$^+$ T cells if they act upon fibroblasts before the cells become infected by CMV. CMV-infected macrophages induce CD8$^+$ cells if they act upon fibroblasts before the cells become infected by CMV. CMV-infected macrophages induce CD8$^+$ T-cell precursors (CTLp) to develop into CMV-specific CD8$^+$ effector cells. Dotted arrows represent spread of CMV virus progeny to uninfected cells. (Reprinted from *Trends Microbiol.,* 6, 190-198, 1998. Hengel, H., Brune, W., and Koszinowski, U.H., Immune evasion by cytomegalovirus. ©1998, with permission from Elsevier Science.)

C. *GAMMAHERPESVIRINAE*

Viruses of this group are able to establish latent infections in lymphoid cells and to induce cellular proliferation. The most important member is the human Epstein-Barr virus (EBV; genus *Lymphocryptovirus*). It infects B lymphocytes and epithelial cells and causes infectious mononucleosis, a benign disease with pharyngitis, lymphadenitis, and sometimes splenomegaly and jaundice. Most infections are asymptomatic. In addition, EBV causes B cell lymphomas in immunocompromised individuals and is associated with Burkitt's lymphoma (Central Africa, New Guinea) and nasopharyngeal carcinoma (South China). More than 90% of adult humans carry latent EBV. The virus persists in B cells as episomal DNA. Murine gammaherpesviruses provide experimental models for the study of virus-host interactions and the exploration of therapeutic strategies. Human herpesvirus 8 is associated with Kaposi's sarcoma, an AIDS-related disease.

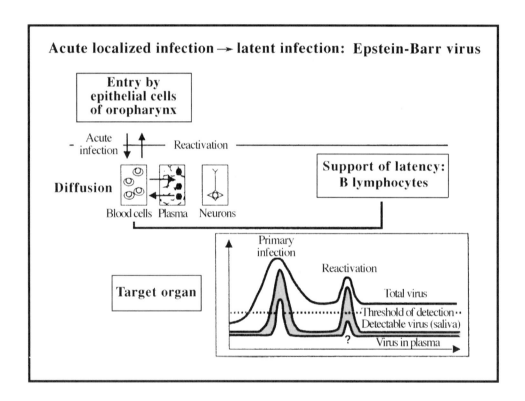

FIGURE 137

Epstein-Barr virus: an example of an acute generalized infection evolving toward latency. The virus persists in B lymphocytes and is therefore detectable during all of latency in circulating infected B cells. During reactivation, free virus can also be detected in saliva and probably the plasma compartment. (Adapted from Maréchal, V., Dehée, A., and Nicolas, J.-C., *Virologie, 1* (special issue), 11, 1997. With permission of John Libbey Eurotext.)

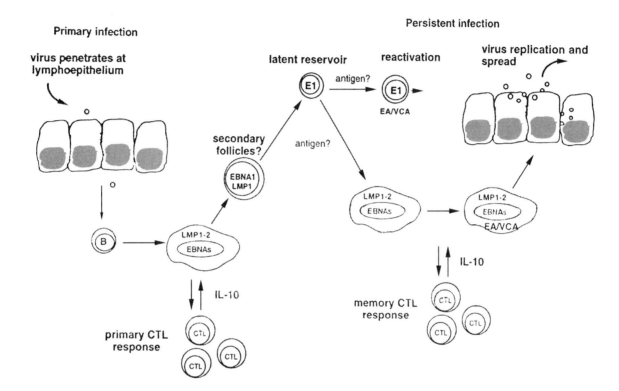

FIGURE 138

Model of lytic and latent infection by EBV. The virus invades lymphoepithelial organs in the oropharynx and infects and transforms B cells. B blast cells are controlled by CTL (cytotoxic lymphocytes), but some enter the resting small B cell pool and form the latent virus reservoir. Upon reactivation, the virus replicates in the oropharynx for spread to new hosts. (From Hill, A.B. and Masucci, M.G., *Sem. Virol.,* 8, 361, 1998. With permission.)

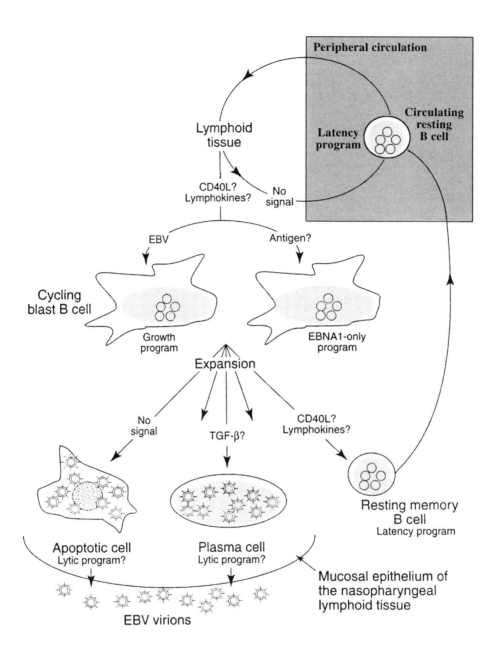

FIGURE 139

Epstein-Barr virus pathogenesis. During acute infection, EBV infects normal B cells and induces a growth program. EBV persists in resting circulating B cells with its genome present in limited copy numbers as covalently closed episomes. It expresses only minimal genetic information (latency program) and does not elicit a cytotoxic cell (CTL) response. On entering lymphoid tissue, the cell may encounter no signals and may pass back into the peripheral circulation, become terminally differentiated, or apoptose. The last two alternatives could lead to viral reactivation (lytic program). In immunosuppressed patients, cells expressing the growth program survive longer because of reduced numbers of CTLs. Longer survival leads to higher numbers and increased risk of secondary genetic effects leading very rare cells to develop into an immunoblastic lymphoma. (Reprinted from *Trends Microbiol.,* 4, 204-208, 1996. Thorley-Lawson, D.A., Miyashita, E.M., and Khan, G., Epstein-Barr virus and the B cell: that's all it takes. © 1996, with permission of Elsevier Science.)

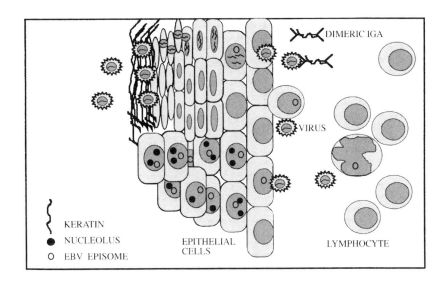

FIGURE 140

Events in the development of nasopharyngeal carcinoma (NPC). EBV reactivation and replication occurs in lymphocytes trafficking in mucosal regions. This virus may induce the expression of EBV-specific dimeric IgA which may facilitate entry into epithelial cells. In the normal course of reactivation, the virus would enter epithelial cells and would replicate as the cells differentiated. In the genesis of NPC, the virus establishes a latent infection and the epithelial cells maintain the EBV episome with expression of critical genes such as LMP1. The latency infected cell is transformed, forming a clone that rapidly develops into invasive carcinoma. (From Raab-Traub, N., *Sem. Virol.,* 7, 315, 1996. With permission.)

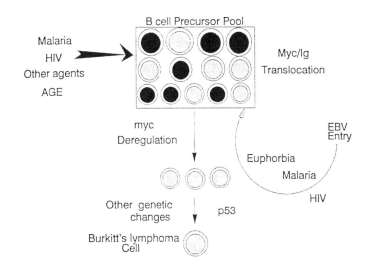

FIGURE 141

Possible events in the pathogenesis of Burkitt's lymphoma. The B cell precursor pool consists of cells at various stages of differentiation indicated by differences in size. EBV may infect a proportion of these cells, as indicated by different aspects of cell nuclei. Environmental factors may modify the site of these different differentiation compartments, which in turn may influence the pattern of breakpoint locations on chromosomes 8 and 14. Age may influence the overall size of the pool and possible subcompartment size. A myc/Ig translocation is probably insufficient to induce overt neoplasia, which is likely to require additional genetic lesions occurring in a single clone. (From Magrath, I., Jain, V., and Bhatia, K., in *The Epstein-Barr Virus and Associated Diseases,* Tursz, T. and coll., Eds., INSERM/John Libbey Eurotext, 1993, 377. With permission of INSERM and John Libbey Eurotext.)

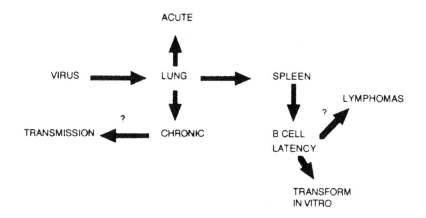

FIGURE 142

Pathways of infection and pathogenesis in murine gammaherpesvirus MHV-68. The virus infects mice in the laboratory, replicating in the lung. It then spreads to the B-cell compartment and causes splenomegaly, a process reminiscent of infectious mononucleosis in humans, becomes latent in B-cells, and persists. Long-term infection is associated with the development of lymphoproliferative disease. CD8 T-cells are key cells in resolving the primary infection and in maintaining the numbers of latently infected B-cells. A critical interplay exists between B-cells and CD4 T cells in the genesis of splenomegaly and the numbers of latently infected B-cells. (From Nash, A.A., Usherwood, E.J., and Stewart, J.P., *Sem. Virol.,* 7,125, 1996. With permission.)

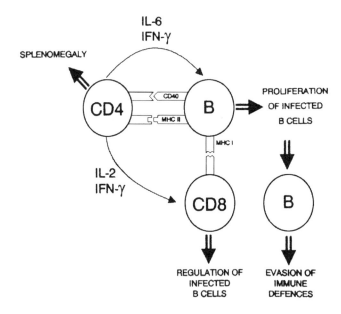

FIGURE 143

In MHV-68, CD4 T-cell/B-cell interaction is necessary for splenomegaly and proliferation of infected B-cells. This can occur via cell contact involving MHC class II and CD40 recognition or via cytokines. B-cells also serve a key role in presentation of virus antigen to CD8 T-cells. It is hypothesized that these cells limit the number of infected B-cells, thereby reducing splenomegaly. Clearly, some latently infected B-cells evade host recognition and establish the pool of latent cells in the host. (From Nash, A.A., Usherwood, E.J., and Stewart, J.P., *Sem. Virol.,* 7, 125, 1996. With permission.)

6.X. ORTHOMYXOVIRIDAE

Linear (-) sense ss RNA, segmented
Helical, enveloped

This family has four small genera. Members infect mammals, birds, and ticks. Particles are usually spherical and 80-120 nm in diameter, and consist of an envelope with glycoprotein spikes and 6-8 nucleoprotein complexes (eight in influenza A and B viruses); filamentous forms are common. Gene reassortment occurs in mixed infections by viruses of different genotypes and results in antigenic shifts followed by outbreaks of epidemics.

The most important human pathogen is influenza A virus, which also infects other mammals (pigs, horses, seals) and birds. Pigs and fowl seem to be the natural reservoir. Human influenza is an acute, sometimes fatal respiratory disease of 6-9 days duration with rather uncharacteristic symptoms (fever, myalgia, headache, cough, nasal discharge), necrosis of the respiratory epithelium, and sometimes a pneumonia prone to bacterial superinfection. The virus multiplies locally within the respiratory mucosa and does not enter the body. Influenza A virus has a high degree of variability. Most variations occur in the surface glycoproteins of the virus, against which protective antibodies are directed. The natural evolution of influenza is largely driven by antibody-mediated selection.

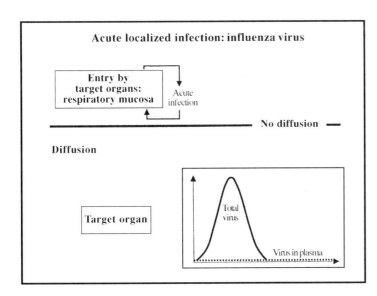

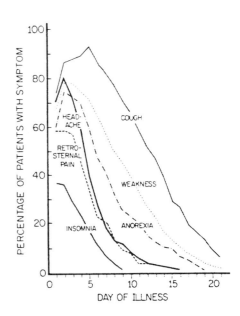

FIGURE 144

Influenza virus: an example of a localized acute infection. Virus replication and infected cells are detected only transitorily at the site of virus entry. (Adapted from Maréchal, V., Dehée, A., and Nicolas, J.-C., *Virologie,* 1 (special issue), 11, 1997. With permission of John Libbey Eurotext.)

FIGURE 145

Frequency and duration of symptoms in 148 patients with influenza A/Hong Kong/68 infection, New South Wales, Australia, 1970. (From Gunton, P.E., *Austr. Family Phys.*, 1, 343, 1972. With permission.)

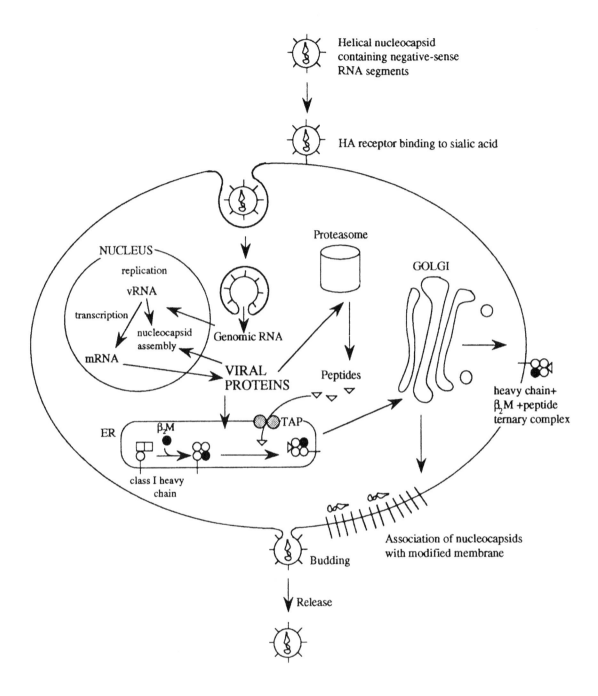

FIGURE 146

Influenza A virus replication and pathway of MHC class I-restricted antigen presentation in an infected cell. Influenza viruses are extremely variable in their surface glycoproteins which induce most of protective antibodies. In contrast, the strong MHC class I-restricted CTL response to infection with virus is essentially specific for internal proteins which are relatively well conserved, and is cross-reactive between different strains of influenza A virus. (From Parker, C.E. and Gould, K. G., *Sem. Virol.*, 7, 61, 1996. With permission.)

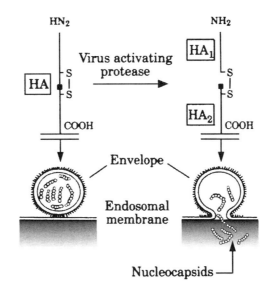

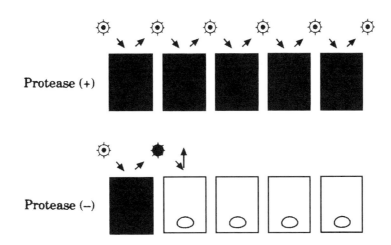

FIGURE 147

Proteolytic activation of influenza virus HA and infectivity. Cleavage activation of HA into disulfide-linked subunits HA1 and HA2 by host proteases is required for influenza virus to express membrane fusion activity and infectivity. Influenza virus can undergo multiple cycles of replication and show pathogenicity in tissues where suitable HA-activating protease(s) is present. (From Tashiro, M. and Rott, R., *Sem. Virol.*, 7, 237, 1996. With permission.)

6.XI. PAPILLOMAVIRIDAE

Linear dsDNA
Cubic, naked

Papillomaviruses were formerly assigned to the family *Papovaviridae,* but now constitute an independent family with a single genus. Particles are icosahedra of 35 nm in diameter. Viruses are very host-specific and occur in large numbers of mammals and birds, causing local benign hyperplasias of the skin (warts) or mucosae (papillomas, condylomas). Papillomaviruses infect squamous epithelial cells of the skin or mucosa. Replication is linked to cell differentiation and complete viruses are found in superficial cell layers only. On occasion, warts or papillomas may progress to malignant tumors (e.g., the Shope papilloma).

There are over 70 types of human papillomaviruses (HPVs). Some are associated with precancerous conditions, notably HPVs 16 and 18, and are consistently found in cervical carcinomas and anogenital cancers. Viral DNA is found in about 90% of these tumors, most tumors contain integrated viral DNA, and most or all HPV-positive cancer biopsies, and all HPV-containing cell lines derived from cervical cancer, reveal specific transcripts originating from two specific open reading frames (E6, E7) of persisting HPV DNA. The DNA of high-risk HPVs 16 and 18 persists as low-copy episomes, but may also be present in integrated form. On the molecular level, the viral E6 and E7 proteins immortalize human keratocytes and abrogate the activities of tumor suppression proteins p53 and pRB.

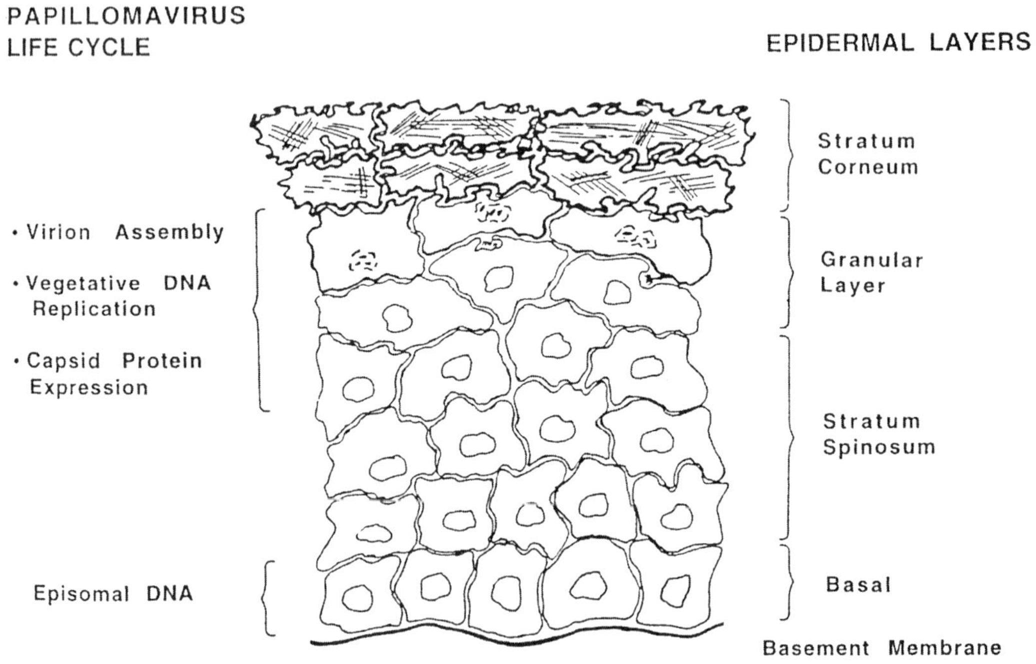

PAPILLOMAVIRUS LIFE CYCLE   **EPIDERMAL LAYERS**

- Virion Assembly
- Vegetative DNA Replication
- Capsid Protein Expression

Episomal DNA

Stratum Corneum
Granular Layer
Stratum Spinosum
Basal
Basement Membrane

FIGURE 148

Stages of infection by papilloma virus. (Drawing by M. Ceniceros and P. Howley, from Ahmed, R., Morrison, L.A., and Knipe, D.M., in *Fields Virology,* 3rd ed., Vol. 1, Fields, B.N., Knipe, D.M., and Howley, P.M., Eds.-in-chief, Lippincott-Raven, Philadelphia, 1996, 219. With permission by P.M. Howley.)

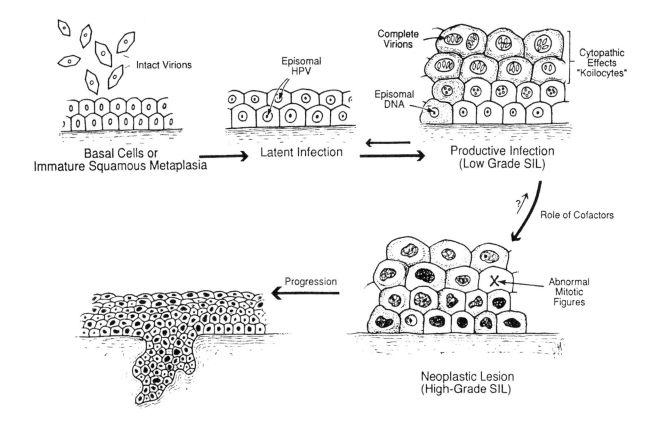

FIGURE 149

Life cycle of HPV. The first step in HPV infection is contact of infectious virus with basal cells or immature squamous metaplastic cells. This can produce either a latent infection or a productive infection. In a latent infection HPV DNA remains as an episome in the nucleus of the infected basal cell. In productive infections, viral replication becomes uncoupled from cellular DNA synthesis and large amounts of viral DNA and proteins are made in the intermediate and superficial cell layers of the epithelium, producing the characteristic cytopathic effects of HPV. During the development of high-grade SIL (squamous intraepithelial lesions) and invasive squamous cancers, additional cellular and viral events take place, resulting in the formation of a "true" cancer precursor. (Fig. 7.3 from Wright, T.C., Kurman, R.J., and Ferenczy, A., in *Blaustein's Pathology of the Female Genital Tract,* Kurman, E., Ed., Springer, New York, 1994, 229. © Springer-Verlag. With permission.)

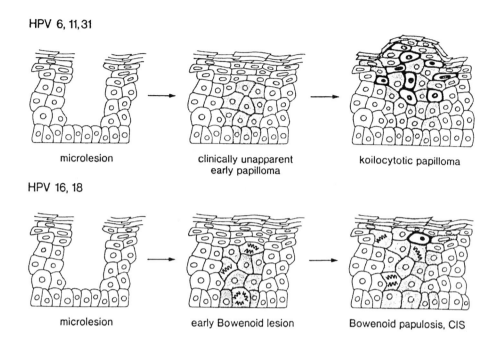

FIGURE 150

Natural history of HPV infections. Viral infections primarily appear to occur via microlesions or in proliferating cells exposed to the surface at the transformation zone. HPV-6, 1, and 31 infections result mostly in koilocytotic dysplasias at cervical sites, whereas HPV-16 and 18 infections lead to atypical mitotic figures, Bowenoid changes, and carcinoma *in situ* (CIS). (From Zur Hausen, H. and Schneider, A., in *The Papovaviridae,* Vol. 2, *The Papillomaviruses,* Salzman, N.P. and Howley, P.M., Eds., Plenum Press, New York, 1987, 245. With permission.)

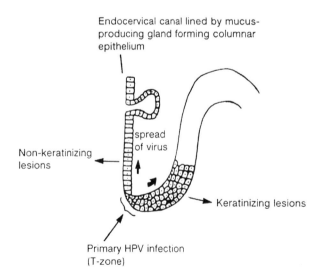

FIGURE 151

Possible sequence of events in HPV infections of the uterine cervix. (Fig. 15 from Koss, L.G., in *Papillomavruses and Human Disease,* Syrjänen, K., Gissmann, L., and Koss, L., Eds., Springer, Berlin, 1982, 235. © Springer-Verlag. With permission.)

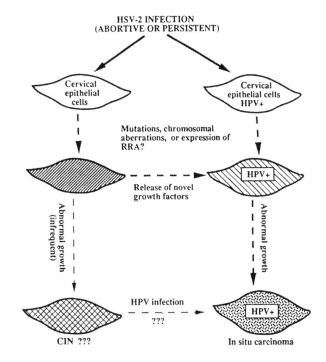

FIGURE 152

Herpes simplex virus 2 (HSV-2) as a possible cofactor in cervical cancer. CIN, cervical intraepithelial neoplasia. (From Jones, C., *Clin. Microbiol. Rev.,* 8, 549, 1995. With permission.)

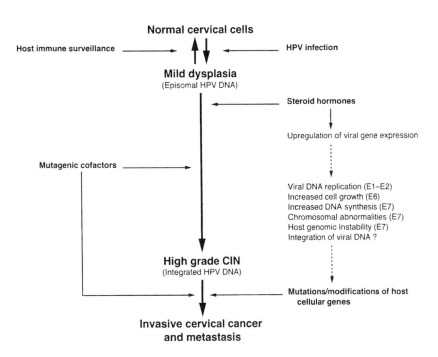

FIGURE 153

Model for the role of steroid hormones in oncogenesis of human-papillomavirus-infected (HPV-infected) cervical tissues and cervical intraepithelial neoplasia (CIN) lesions. E1, E2, E6, E7, viral genes. (Reprinted from *Trends Microbiol.*, 2, 229-234, 1994. Pater, M.M., Mittal, R., and Pater, A., Role of steroid hormones in potentiating transformation of cervical cells by human papillomaviruses. © 1994, with permission of Elsevier Science.)

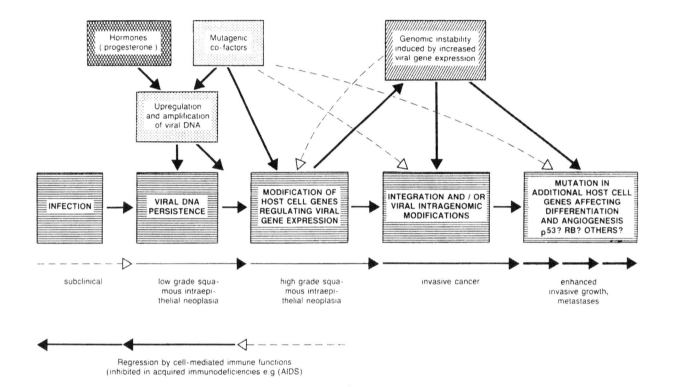

FIGURE 154

Pathogenesis of anogenital cancer. Substantial evidence indicates an etiological role of specific human papilloma viruses in anogenital cancer and its malignant precursors. In addition to viral infection and viral gene expression, modifications of host cell genes appear to be required for malignant progression of infected cells. The expression of viral oncoproteins in cells infected by "high-risk" types (e.g., HPV 16, HPV 18), in contrast to low-risk types (e.g., HPV 6, HPV 11), results in chromosomal instability and apparently in accumulation of mutational events. These "endogenous" modifications seem to be most important in the pathogenesis of premalignant lesions and tumor progression. Exogenous mutagens could act as additional factors. (From Zur Hausen, H., *Virology,* 184, 9, 1991. With permission.)

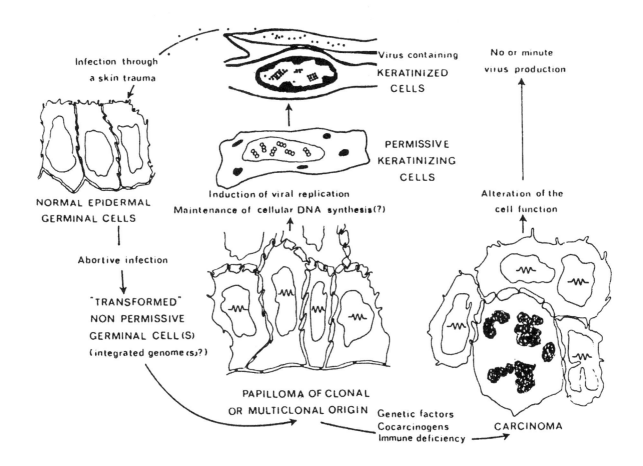

FIGURE 155

Model of Shope papilloma virus interaction with cottontail epidermal cells. (From Orth, G., Breitburd, F., Favre, M., and Croissant, O., in *Origins of Human Cancer*. B. *Mechanisms of Carcinogenesis*, Hiatt, H.H., Watson, J.D., and Winsten, J.A., Eds., Cold Spring Harbor Laboratory Press, Cold Spring Harbor, 1977, 1043. With permission.)

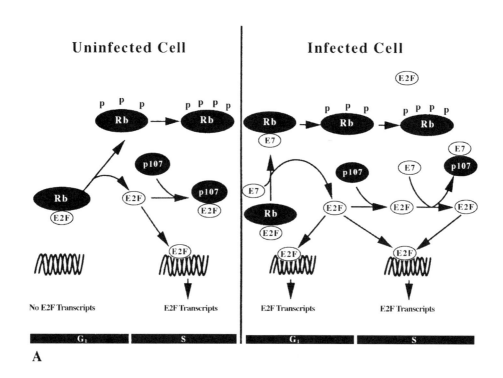

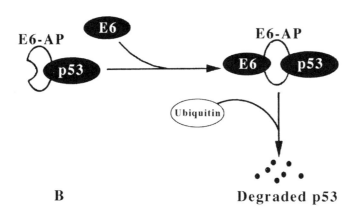

FIGURE 156

Cellular proteins involved in papillomavirus-induced transformations. (A) In uninfected cells the activity of transcription factor E2F appears to cause the cell entering S-phase. In resting cells its activity is regulated to some extent by binding to the retinoblastoma protein, Rb. At the G_1/S transition, Rb becomes more phosphorylated and no longer binds E2F, thus allowing transcription of E2F-responsive genes and entry into S-phase. Rb becomes still more phosphorylated and the activity of E2F during S-phase appears to be regulated by the Rb-related protein p107. In cells infected with high-risk HPV's, the early protein, E7, has an affinity for phosphorylated Rb and will bind to it causing the release of unbound E2F. The formal recognition of E2F during G_1 is lost and the cells may enter S phase prematurely. E7 also appears to have affinity for p107 and binds to it disrupting the control during the S phase.

(B) HPV early protein, E6, from the high-risk viruses, causes the degradation of p53. This degradation appears to occur via the ubiquitin pathway. E6 and the E6-associated protein (E6-AP) are found in a complex with p53 in HPV-infected cells. It has been suggested that the E6-AP may be homologous to one of the proteins in the ubiquitin pathway and that p53 in the complex thus becomes targeted for degradation. (Adapted from Swan, D.C., Vernon, S.D., and Icenogle, J.P., *Arch. Virol.*, 138, 105, 1994. © Springer-Verlag. With permission.)

6.XII. PARAMYXOVIRIDAE

Linear (-) sense ssRNA, segmented
Enveloped, helical

Viruses infect mammals, birds, and reptiles. They are classified into two subfamilies and five genera and comprise major human and animal pathogens. The subfamily *Paramyxovirinae* has three genera represented by Sendai, measles, and mumps viruses, respectively. Particles are roughly spherical and 150-300 nm in size, and consist of an envelope with spikes and a single nucleocapsid. Filamentous forms are common.

MEASLES
Measles virus and its relatives, which include important animal pathogens such as rinderpest and canine distemper virus, constitute the genus *Morbillivirus*. Measles is an exclusively human, highly contagious childhood disease and a major cause of child mortality in developing countries. The disease is characterized by fever, conjunctivitis, and a maculopapular rash. Complications are frequent and include otitis media, pneumonia, and encephalomyelitis. The dreaded inclusion body encephalitis and subacute sclerosing panencephalitis (SSPE) develop after months and even years after infection and are characterized by virus persistence in neuronal and glial cells and absence of free virus.

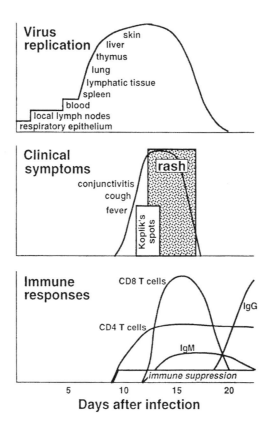

FIGURE 157
Pathogenesis of measles. Virus replication begins in the respiratory epithelium and spreads to monocyte/macrophages, endothelial cells, and epithelial cells in the blood, thymus, spleen, lymph nodes, liver, skin, and lungs and to the conjuctivae and the mucosal surfaces of the gastrointestinal, respiratory, and genitourinary tracts. The rash appears at the time of the virus-specific immune response. Clearance of the virus is approximately coincident with fading of the rash. (From Griffin, D.E. and Bellini, W.J., in *Fields Virology*, 3rd ed., Vol. 1, Fields, B.N., Knipe, D.M., and Howley, P.M., Eds.-in-chief, Lippincott-Raven, Philadelphia, 1996, 1267-1312. © Lippincott Williams & Wilkins.)

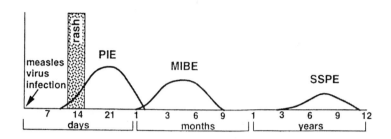

FIGURE 158

Occurrence of the three neurological complications of measles. PIE, Postinfectious encephalomyelitis; MIBE, measles inclusion body encephalitis; SSPE, subacute sclerosing panencephalitis. (From Griffin, D.E. and Bellini, W.J., in *Fields Virology*, 3rd ed., Vol. 1, Fields, B.N., Knipe, D.M., and Howley, P.M., Eds.-in-chief, Lippincott-Raven, Philadelphia, 1996, 1267-1312. © Lippincott Williams & Wilkins. With permission.)

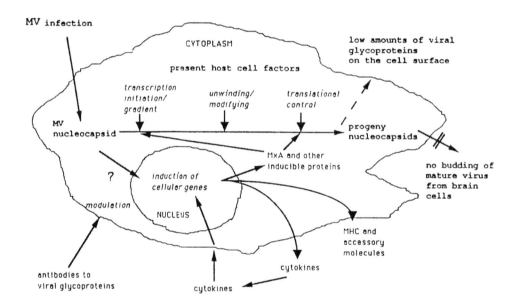

FIGURE 159

Host factors influencing Measles Virus (MV) gene expression in neural cells. Host factors and their mode of action contributing to the establishment of persistent CNS infections by modulating viral and cellular gene functions are summarized here. Transcriptional efficiency of viral RNA is downregulated by cellular components present in the cell or induced by cytokines (as MxA) or antiviral antibodies. MV-specific RNA may be occasionally modified by the cellular unwindase present in the cytoplasm of brain cells. Viral mRNAs, either modified or unmodified, are barely translated into structural proteins, in particular in differentiated brain cells, ending up in prolonged, inefficient replication of the viral genome and low expression of envelope proteins. Consequently, budding of infectious virus is delayed or may even be absent. As a second event, cellular gene expression is altered upon MV infection, as indicated by expression of surface molecules and the release of a characteristic set of cytokines. These cytokines may enhance not only the antiviral immune response but also directly control of viral gene expression as shown for the type I interferon-induced MxA protein. (Fig 2 from Schneider-Schaulies, S., Schneider-Schaulies, J., Dunster, L.M., and Ter Meulen, V., *Curr. Topics Microbiol. Immunol.*, 191, 101, 1995. © Springer-Verlag. With permission.)

MUMPS

Mumps virus is a member of the genus *Rubulavirus*. Mumps is a common, benign disease of children and young adults and is characterized by an inflammation of the salivary glands, especially the parotids. The most frequent complications are orchitis and benign meningoencephalitis. Ovaries, pancreas, and thyroid may also be affected.

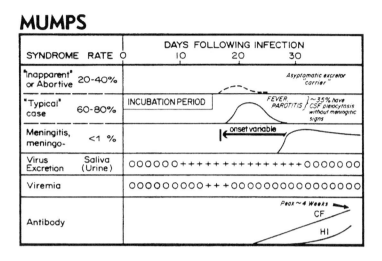

FIGURE 160

General features of mumps. Relative frequency of the forms of clinical expression; virus excretion; antibody response to infection. (Reprinted with permission from Monath, T.P., in *Virology and Rickettsiology*, Vol. I, Part 2, Hsiung, G.-D and Green, R.H., Eds., CRC Press, Boca Raton, 1978, 261. Copyright CRC Press, Boca Raton, Florida.)

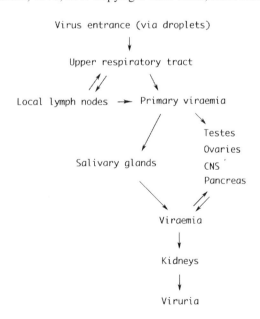

FIGURE 161

Pathogenesis of mumps infection. Primary and secondary viremia can both lead to infection of other target organs than salivary glands. (From Leinikki, P., in *Principles and Practice of Clinical Virology,* 2nd ed., Zuckerman, A.J., Banatvala, J.E., and Pattison, J.R., Eds., © John Wiley & Sons, Chichester, 1990, 375. Reproduced by permission of John Wiley & Sons.)

RESPIRATORY SYNCYTIAL VIRUS (RSV) BRONCHIOLITIS

RSV is a member of the *Pneumovirinae* subfamily which also includes viruses of cattle, mice, and turkeys. Human RSV causes bronchiolitis and bronchitis in infants during the first year of life (and in the elderly). The disease is most severe between the first 2 to 4 months and may be fatal. Symptoms include dyspnea and cyanosis with or without wheezing and emphysema. The development of a vaccine is highly desirable, but some trial vaccines have actually increased the severity of the infection. RSV bronchiolitis may be mediated by cytokines produced by type 2 T helper cells.

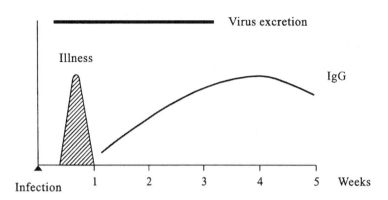

FIGURE 162

Course of respiratory syncytial virus (RSV) infections. (By Ackermann.)

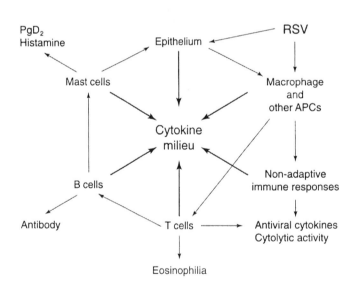

FIGURE 163

All cells in the cascade of immune events triggered by respiratory syncytial virus (RSV) infection contribute to the cytokine milieu and so have a regulatory role in subsequent immune events. The composite effect can bias T helper cell differentiation and determine the phenotype of precursors responsible for virus clearance and the magnitude of immunopathology. PgD$_2$, prostaglandin D$_2$; APCs, antigen presenting cells. (Reprinted from *Trends Microbiol.*, 4, 290-293, 1996. Graham, B.S., Immunological determinants of disease caused by respiratory syncytial viruses. © 1996, with permission of Elsevier Science.)

6.XIII. PARVOVIRIDAE

Linear ssDNA
Cubic, naked

Members of this family infect vertebrates (subfamily *Parvovirinae*, three genera) and insects (*Densovirinae,* three genera). Particles are icosahedra of 18-25 nm in diameter and seem to be the smallest of all viruses. Members of the genus *Dependovirus* (vertebrates) require the presence of adenoviruses or herpesviruses as helpers in replication. Vertebrate parvoviruses are highly tissue-specific and can induce the death of tumor cells. Parvoviruses infect a wide variety of mammals and birds and cause, among others, Aleutian mink disease. B19 virus, a member of the genus *Erythrovirus,* causes aplastic anemia and arthritis in children and adults. Parvovirus DNA may integrate into the host genome. The virus replicates in erythrocyte precursors. Infection during pregnancy may result in fetal hydrops and death or in persistent infection with red cell aplasia.

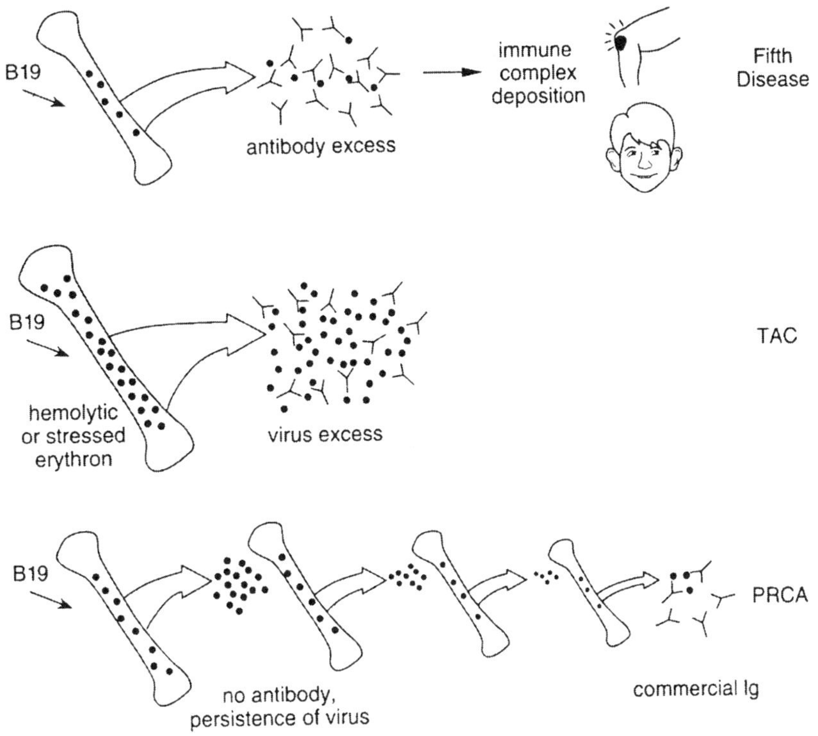

FIGURE 164

Pathogenesis of diseases caused by B19 parvovirus. (A) Infection of children and adults.. PRCA, pure red cell aplasia; TAC, transient aplastic crisis. (From Young, N.S., in *Fields Virology*, 3rd ed., Vol. 2, Fields, B.N., Knipe, D.M., and Howley, P.M., Eds.-in-chief, Lippincott-Raven, Philadelphia, 1996, 2199-2220. © Lippincott Williams & Wilkins. With permission.)

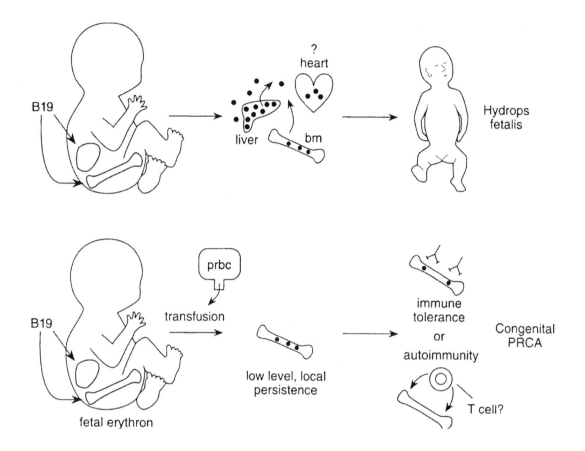

FIGURE 165

Pathogenesis of diseases caused by B19 parvovirus. (B) Diseases resulting from fetal infection. Bm, bone marrow; PRCA, pure red cell aplasia; prbc, purified red blood cells. (From Young, N.S., in *Fields Virology*, 3rd ed., Vol. 2, Fields, B.N., Knipe, D.M., and Howley, P.M., Eds.-in-chief, Lippincott-Raven, Philadelphia, 1996, 2199-2220. © Lippincott Williams & Wilkins. With permission.)

6.XIV. PICORNAVIRIDAE

(+) sense ssRNA, nonsegmented
Cubic, naked

This family consists of nine genera of mammalian and bird viruses and includes major human and animal pathogens. Most picornaviruses are specific to one or a very few host species. Particles are icosahedra of 28-30 nm in diameter. Transmission is fecal-oral or by air or fomites. Infections are acute. Viruses are very resistant against the environment. The most important animal disease, harmless in humans and not illustrated here, is foot-and-mouth-disease of cattle and other cloven-hoofed animals (genus *Aphthovirus*). The genus *Enterovirus* comprises about 100 serotypes, including polioviruses (three serotypes) and agents that are traditionally designated "Coxsackie virus" and "ECHO virus". Particles are resistant to the acid pH in the stomach (pH 2). Viruses mainly replicate in the gastrointestinal tract, but also multiply in other locations (nerves, muscle). Infections are asymptomatic or cause a wide range of disease from common cold to myocarditis and paralysis.

POLIOMYELITIS

This disease, now targeted for global eradication, was once one of the most-feared viral infections. Humans are the most-feared main target and reservoir, although chimpanzees in the wild may be infected. The disease has four forms: inapparent, abortive (fever, headache, malaise, sore throat), meningitis without paralysis, or paralytic poliomyelitis. The latter may result in permanent flaccid paralysis or be fatal. After ingestion, the virus multiplies in the throat and intestine and then reaches regional lymph nodes and the blood stream (viremic stage). Its final and principal targets are the motor neurons of the spinal cord, which may be destroyed. Motor cells of the brain and many other types of brain cells may also be affected. The post-polio syndrome, which appears 30 to 40 years after paralytic poliomyelitis, is characterized by muscle atrophy and slow progressive paralysis. It is attributed to chronic overuse of surviving motor units.

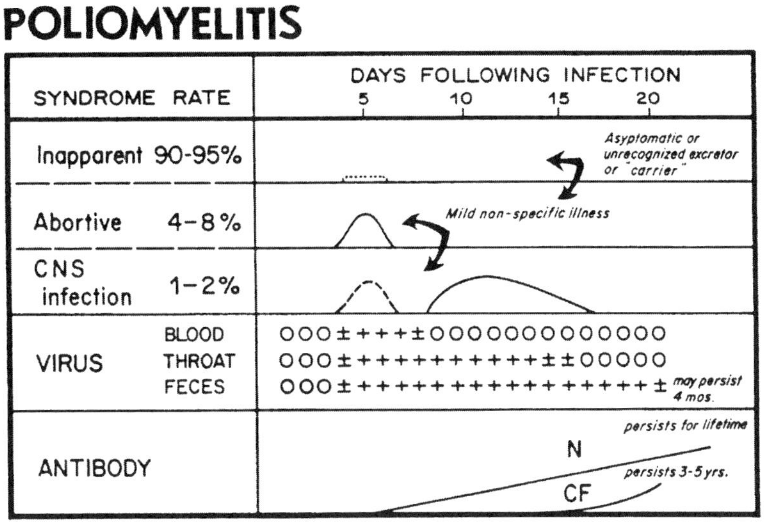

FIGURE 166

General features of poliomyelitis. Relative frequency of the forms of clinical expression; virus excretion; antibody response to infection. (Reprinted with permission from Monath, T.P., in *Virology and Rickettsiology*, Vol. I, Part 2, Hsiung, G.-D and Green, R.H., Eds., CRC Press, Boca Raton, 1978, 261. Copyright CRC Press, Boca Raton, Florida.)

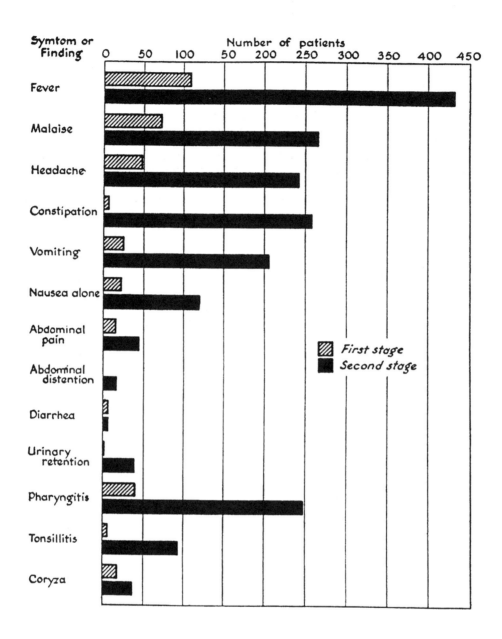

FIGURE 167

Incidence of various signs and symptoms of acute poliomyelitis as observed during the so-called prodromal illness (163 cases) as well as during the preparalytic stage in a series of 464 children treated at the University of Minnesota hospitals in 1946. (From Grulee, C. and Panos, T.C., *Am. J. Dis. Children*, 75, 24, 1948. Copyright 1948, American Medical Association.)

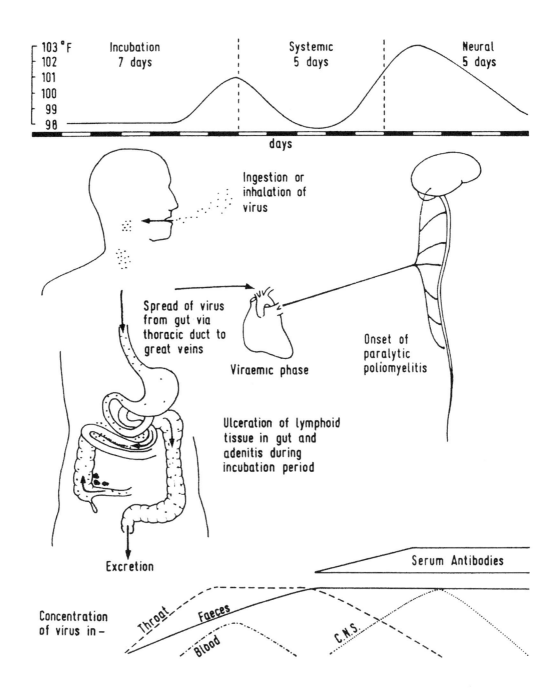

FIGURE 168

Clinical course and pathogenesis of poliomyelitis. (From Swain, R.H.A. and Dodds, T.C., *Clinical Virology,* E. & S. Livingstone, Edinburgh, 1967, 167. With permission.)

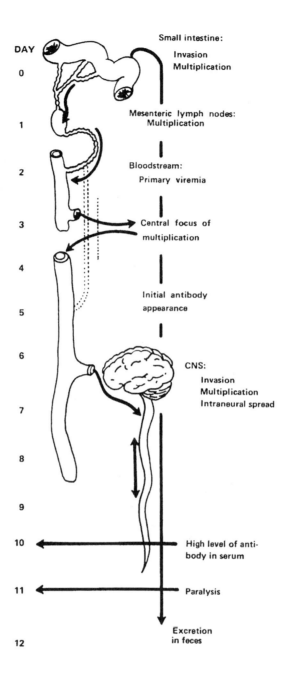

FIGURE 169

Pathogenesis of poliomyelitis. Virus enters by way of the alimentary tract and multiplies locally at the initial sites of virus implantation (tonsils, Peyer's patches) or the lymph nodes that drain these tissues, and virus begins to appear in the throat and feces. Secondary virus spread occurs by way of the bloodstream to other susceptible tissues, namely, other lymph nodes, brown fat, and the CNS. Within the CNS the virus spreads along nerve fibers. If a high level of multiplication occurs as the virus spreads through the CNS, motor neurons are destroyed and paralysis occurs. The shedding of virus into the environment does not depend on secondary virus spread to the CNS. (Derived from Fenner, F., *Med. J. Aust.,* i, 205, 1956; from Melnick, J.L., in *Fields Virology*, 3rd ed., Vol. 1, Fields, B.N., Knipe, D.M., and Howley, P.M., Eds.-in-chief, Lippincott-Raven, Philadelphia, 1996, 655-712. © Lippincott Williams & Wilkins. With permission.)

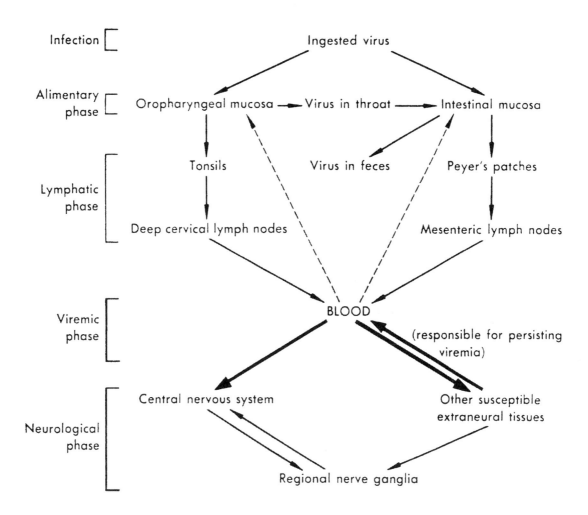

FIGURE 170

Pathogenesis of poliomyelitis. Model is based on data obtained in man and chimpanzees. (Based on Bodian, D., *Science,* 122, 105, 1955; and Sabin, A.B., *Science,* 123, 1151, 1956; from Ginsberg, H.S., in *Microbiology*, 3rd ed., Davis, B.D., Dulbecco, R., Eisen, H.N., and Ginsberg, H.S., Eds. Harper & Row, Hagerstown, 1980, 1095-1117. © 1955 and 1956 American Association for the Advancement of Science, © 1980 Lippincott Williams & Wilkins. With permission.)

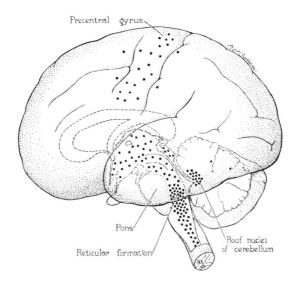

FIGURE 171

Lateral view of human brain, with schematic transparent projection of the midsagittal surface of the brainstem. General distribution of lesions of poliomyelitis is indicated by large dots. Lesions in the cerebral cortex are largely restricted to the precentral gyrus, and those in the cerebellum of the roof nuclei. Lesions are generally found widespread in the brainstem centers, with a number of striking exceptions, such as the nuclei of the basis pontis, and the inferior olivary nuclei. (Reprinted from *American Journal of Medicine,* 6, 563-578. Bodian, D., "Histopathological basis of clinical findings in poliomyelitis," © 1949, with permission of Excerpta Medica Inc.)

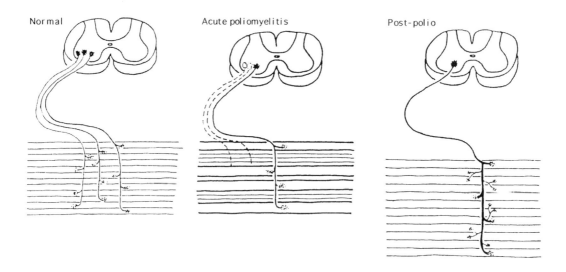

FIGURE 172

Representation of motor units to a muscle. *Normal* represents the 100 to 1000 motor neurons of a muscle and the 5 to 1500 muscle fibers each axon innervates. *Acute poliomyelitis* depicts viral destruction of some of the anterior horn cells with atrophy of denervated muscle fibers. *Postpolio* represents sprouting by recovered nerve cells with reinnervation of the orphaned muscle fiber and subsequent hypertrophy. (From Smith, L.K. and Mabry, M., in *Neurological Rehabilitation,* 3rd. ed., Umphred, D.A., Ed., Mosby, St. Louis, 1995, 571. With permission.)

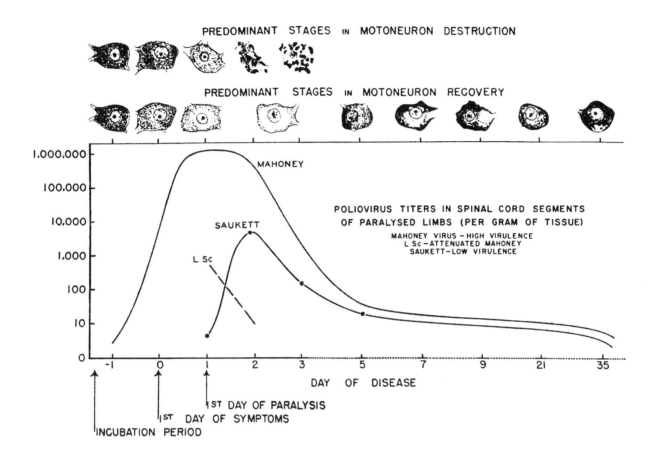

FIGURE 173

Sequence of cytopathological changes in motoneurons in the course of destruction, and those chromatolysed, but able to recover. The approximate time course of changes is shown by parallel curves that show the rise and decline of viral concentration in the rhesus monkey spinal cord. Peak levels of virus concentration are attained at the time when the predominant stage of cell change in the motoneuron population is that of diffuse cytoplasmic chromatolysis. Note particularly the low levels of infectivity of the Saukett strain on the first day of the paralytic signs, even in the segments of the spinal cord that are associated with severely paralyzed extremities. The Mahoney and Saukett strains are tissue-culture-adapted strains that are used in the routine production of vaccine. The L Sc strain is a tissue-adapted sample that was obtained through the courtesy of A.B. Sabin. (From Bodian, D., in *Conference on Cellular Biology, Nucleic Acids, and Viruses,* v. St. Whitelock, O., Ed., N.Y. Acad. Sci., Special Publication No. V, 57, 1957. With permission.)

ENTEROVIRUS MYOCARDITIS

Coxsackie B serotypes, and less frequently ECHO viruses, may cause rare and serious myocarditis and pericarditis. The disease is particularly lethal in newborns and is a leading cause of heart transplants in adults. In the heart, viral lysis of infected myocytes limits viral production, whereas nonspecific (interferon and NK cells) and specific defenses (T cells and antibodies) eliminate infected cells. On the other hand, infections can induce autoimmune responses with lysis of myocytes by heart-specific antibodies.

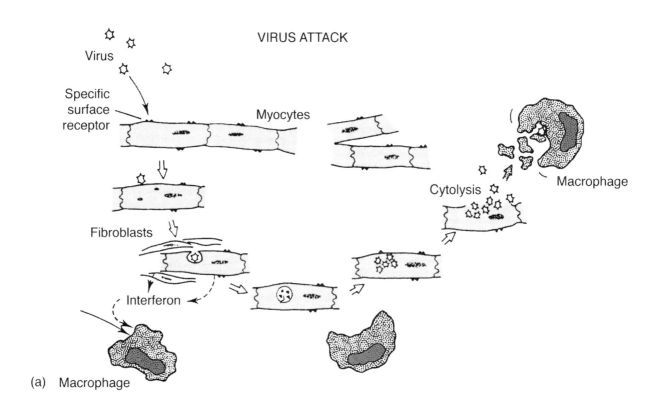

FIGURE 174

Coxsackie virus infection of the heart. Virus attaches to specific receptors on myocytes, is internalized into lysosomes and the viral protein is enzymatically degraded to release the viral RNA into the cytoplasm. Viral RNA is translated and transcribed in the cytoplasm to produce new progeny virus. The progeny are released by lysis of the infected cell. Infection stimulates the release of interferon (α, β) which increases resistance of surrounding cells to infection and activates both macrophage and natural killer cells. (From Huber, S.A., in *Viral Infections of the Heart*, Banatvala, J.E., Ed., Arnold, London, 1993, 82. With permission.)

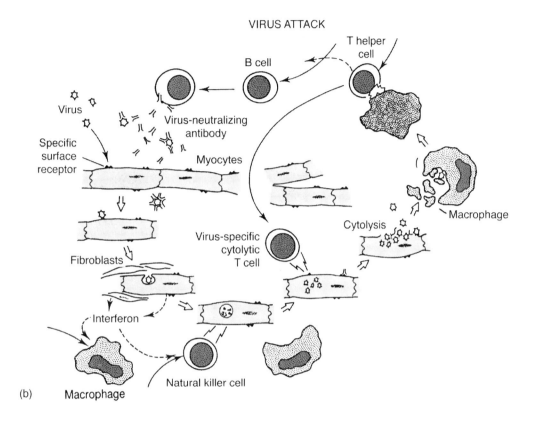

FIGURE 175

Host defense mechanisms in virus infections of the heart. Activated macrophage phagocytize cellular debris and virus and present viral antigens to T and B lymphocytes stimulating both cellular and humoral virus-specific immunity. Virus-specific antibodies bind to virus preventing attachment to uninfected myocytes (limits spread of infection). Virus-specific T and natural killer cells lyse infected cells before completion of virus replication cycle to limit virus production. (From Huber, S.A., in *Viral Infections of the Heart*, Banatvala, J.E., Ed., Arnold, London, 1993, 82. With permission.)

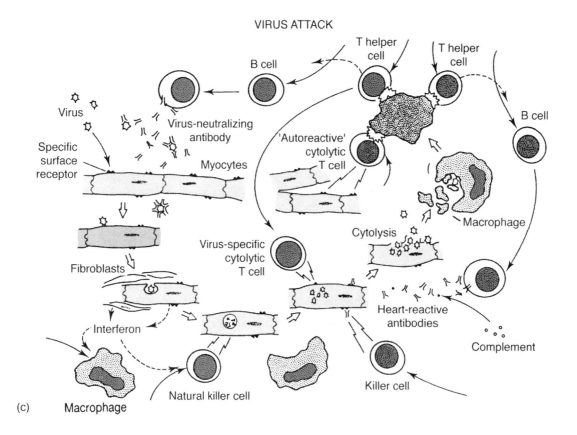

FIGURE 176

Autoimmunity in viral heart disease. Activated macrophage may stimulate autoimmunity by (1) presenting virus antigens which cross-react with host cell molecules (antigenic mimicry) or (2) presenting cellular antigens engulfed during phagocytosis of virus. Both cellular and humoral autoimmunity occur. Autoimmune cytolytic T cells directly lyse myocytes. Heart reactive autoantibodies may lyse myocytes through complement-dependent or antibody-dependent cytotoxicity (ADCC) mechanisms. (From Huber, S.A., in *Viral Infections of the Heart*, Banatvala, J.E., Ed., Arnold, London, 1993, 82. With permission.)

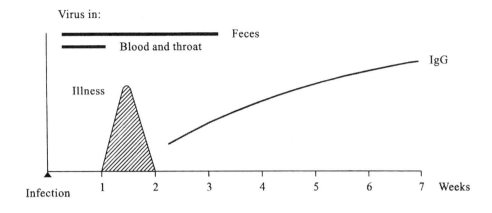

FIGURE 177

Course of a typical enterovirus infection. (By Ackermann.).

COMMON COLD

This syndrome, familiar to everyone and needing no explanations, is caused by a large number of respiratory viruses including about 100 members of the *Rhinovirus* genus. Each of them induces specific immunity. Contrary to enteroviruses, rhinoviruses are inactivated at pH 3.

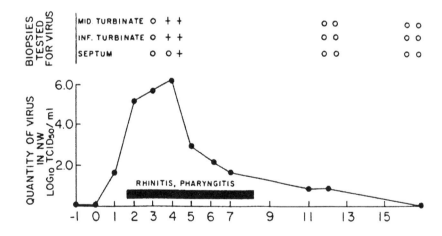

FIGURE 178

Case report of a volunteer inoculated by the intranasal route with rhinovirus type 13. Note the relationship among positive nasal biopsy specimens, quantity of virus in secretions, and occurrence of illness. NW, nose washings; TCID, tissue culture infectious dose. (From Douglas, R.G., Alford, B.R., and Couch, R.B., *Antimicrob. Ag. Chemother.*, 18, 340, 1968. With permission.)

HEPATITIS A

Human and simian hepatitis A viruses form the genus *Hepatovirus*. Clinical symptoms are identical to those of hepatitis B (see above), but the incubation period is relatively short (10-50 days). The infection route is fecal-oral. There are rare cases of fulminant hepatitis, but no tendency to evolve toward chronicity and cirrhosis and liver cancer.

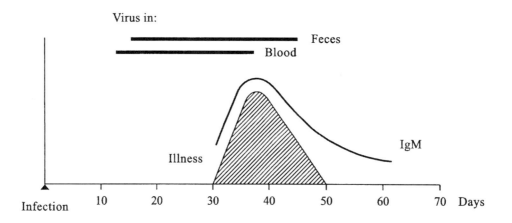

FIGURE 179

The course of hepatitis A virus infection. (By Ackermann.)

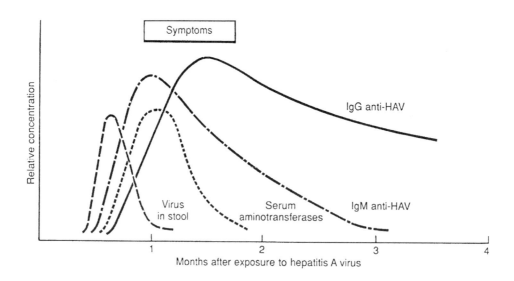

FIGURE 180

Typical course of acute hepatitis A virus (HAV) infection. (From Straus, S.E., in *Mechanisms of Microbial Disease,* 2nd ed. Schaechter, M.E., Medoff, G., and Eisenstein, B.I., Eds., Williams & Wilkins, Baltimore, 1993, 518-530. © Lippincott Williams & Wilkins. With permission.)

6.XV. POXVIRIDAE

Linear dsDNA
Helical, enveloped

This very large virus family is found in mammals, birds (subfamily *Chordopoxvirinae,* eight genera), and insects (*Entomopoxvirinae,* three genera). Particles are brick-shaped or ovoid, measure 220-450 x 140-260 x 110-200 mm, and show considerable structural variation. Chordopoxviruses consist of an envelope, a lipid layer, two lateral bodies, and a helical core. Most vertebrate poxviruses have narrow host ranges and many produce maculopapular or vesicular rashes after systemic or localized infection. Mousepox virus, the agent of ectromelia, is a member of the *Orthopoxvirus* genus. Myxoma virus (*Leporipoxvirus* genus) causes localized benign tumors or severe generalized infection in rabbits. Orf virus, the agent of contagious pustular dermatitis in sheep and goats, is a member of the *Parapox* genus. Poxviruses have evolved several strategies to evade the host immune response and clearance from immunocompetent hosts. Poxviruses are currently investigated for their ability to code for proteins that modulate the activity of cytokines during infection, particularly for those that block interferon-induced antiviral proteins and apoptosis. An example of the accumulating data is given in Fig. 184.

SMALLPOX

The first and only disease to be eradicated from Earth, smallpox was, particularly in the 18th century, one of the great scourges of mankind. Smallpox is characterized by high fever, virus multiplication in the oropharynx, and a maculopapular rash that evolves into vesicles, pustules, and finally scabs and scars. The disease was a major cause of blindness in newborns; lethality was generally about 15%.

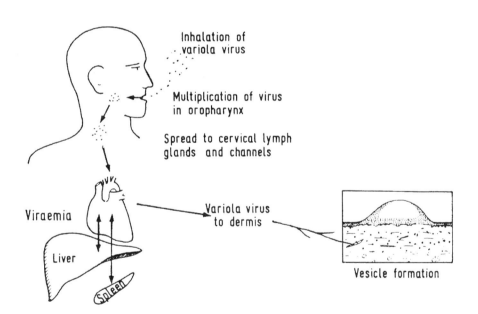

FIGURE 181

Pathogenesis of smallpox. After inhalation, the variola virus multiplies, probably in lymphoid tissue in the oropharynx. A primary viremia is thought to ensue during which the virus is conveyed to the cells of the reticulo-endothelial system. During a short period of clinical improvement the virus rapidly multiplies in this situation until a phase of secondary and massive viremia ensues. Finally the virus multiplies in the cells of the epidermis to produce the characteristic rash. (From Swain, R.H.A. and Dodds, T.C., *Clinical Virology,* E. & S. Livingstone, Edinburgh, 1967, 101. With permission.)

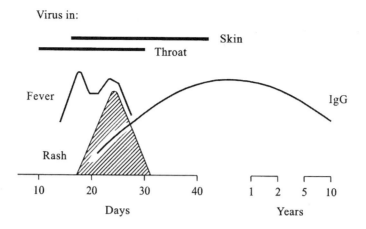

FIGURE 182

Course of smallpox infection. Note the biphasic temperature. (By Ackermann.)

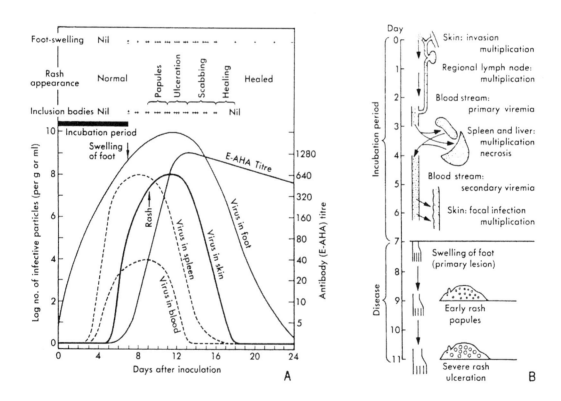

FIGURE 183

(A) Growth curves of virus in foot, spleen, blood, and skin of mice inoculated in the footpad with a small dose of "Moscow" ectromelia virus. Development and disappearance of primary lesion and rash are shown, as is the occurrence of inclusion bodies in skin in sections stained with Mann's stain. (B) Pathogenesis of mousepox. E-AHA, ectromelia antihemagglutinin. (Modified from Fenner, F., *Lancet*, ii, 915, 1948 © by the Lancet Ltd. With permission.)

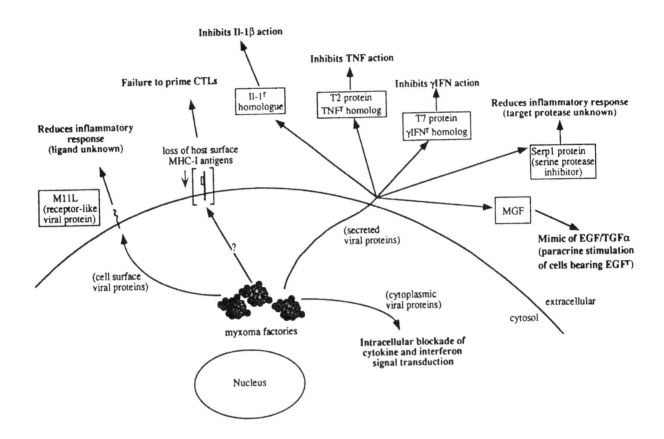

FIGURE 184

Strategies of immune modulation by myxoma virus. Three classes of myxoma virus proteins (secreted, cell surface, and intracellular), produced in cytoplasmic virus factories, are illustrated. Mechanisms of immunosuppression and cytokine/growth factor modulations utilized by the viral proteins shown in boxes are described in greater detail in the original paper. (From McFadden, G. and Graham, K., *Sem. Virol.*, 5, 421, 1994. With permission.)

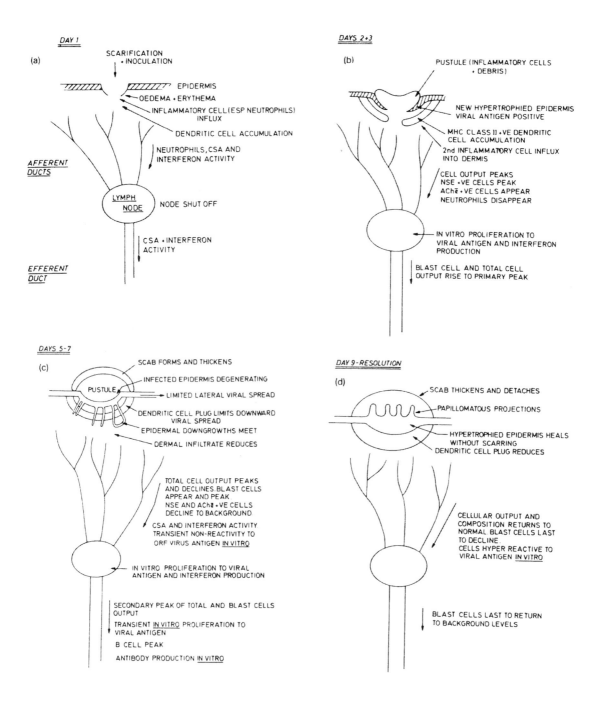

FIGURE 185

Kinetics of local immune responses generated to orf virus infection in sheep: (a) one day, (b), 2-3 days, (c) 5-7 days and (d) 9 days after inoculation. AChe, acetycholine esterase; CSA, colony-stimulating activity; NSE, nonspecific esterase; Ve, dendritic cells. (Reprinted from *FEMS Immunol. Med. Microbiol.*, 8, Yirrell, D.L., Norval, M., and Reid, H.W., Local epidermal virus infections: comparative aspects of vaccinia virus, herpes simplex virus and human papillomavirus in man and orf virus in sheep, 1-12. © 1994, with permission from Elsevier Science.)

6.XVI. REOVIRIDAE

Linear segmented dsRNA
Cubic, naked

This large family has nine genera. Members infect mammals, birds, fish, shellfish, insects, ticks, crustaceans, and plants. Genera differ in host range, capsid structure, number of RNA molecules, and modes of virus entry and egress. Particles are icosahedra of 60-80 nm in diameter and consist of one or two capsids surrounding an inner core with 10-12 RNA molecules. Five genera comprise vertebrate viruses. Members of the genus *Rotavirus* cause diarrheas in infants and young animals and are a major cause of child mortality in developing countries. The genus *Orthoreovirus* seems to be the only reovirus group whose pathogenesis has been illustrated in diagrams. Orthoreoviruses infect mammals (horse, cattle, sheep, swine, mice) and birds; human infections are rare. In suckling mice, the virus is transmitted through synapses rather than from cell to cell. Orthoreoviruses are therefore good markers for the identification of neuronal pathways. They may persist in the CNS because of its relative isolation from the immune system.

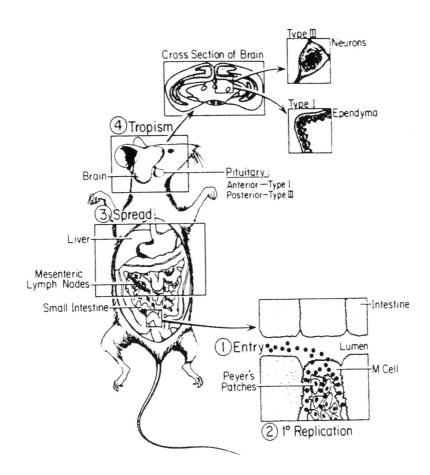

FIGURE 186

Stages in the pathogenesis of reovirus infection. In natural infection, reovirus enters the mouse host through the gastrointestinal tract, penetrates the intestinal epithelium through Peyer's patches, and, under certain circumstances, spreads to the brain. Type 1 virus is localized in ependymal cells, and type 3 virus in neuronal cells. (From Sharpe, A.H. and Fields, B.N., *New England J. Med.,* 312, 486, 1985. © 1985 Massachusetts Medical Society. All rights reserved.)

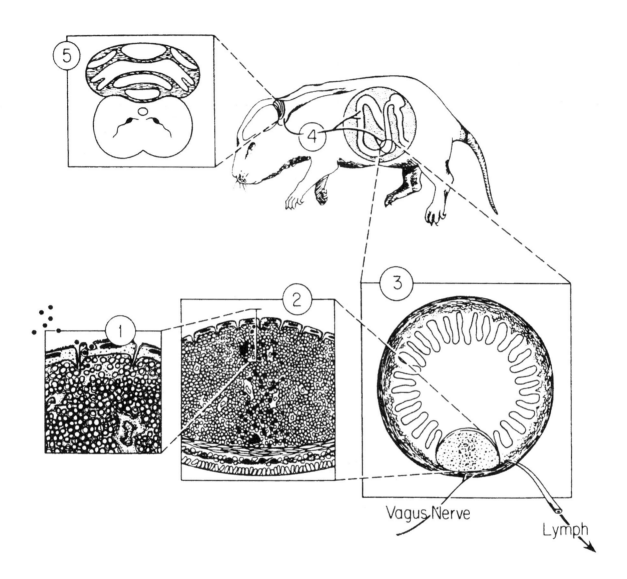

FIGURE 187

Spread of serotype 3 from the intestinal lumen to the CNS. (1) Reovirus (and poliovirus) particles transcytosed by M cells overlying ileal Peyer's patches. (2) Reovirus replication in mononuclear cells in the Peyer's patch and in adjacent myenteric neurons between the muscle layers between the patch. (3) Entry from Peyer's patches into efferent lymphatic capillaries and nerves. (4) Spread from the intestine to the CNS via vagus nerve fibers. (5) Initial infection in the CNS in neurons of the dorsal motor nucleus of the vagus nerve in the brain stem. (From Morrison, L.A. and Fields, B.N., *J. Virol.,* 65, 2767, 1991. With permission.)

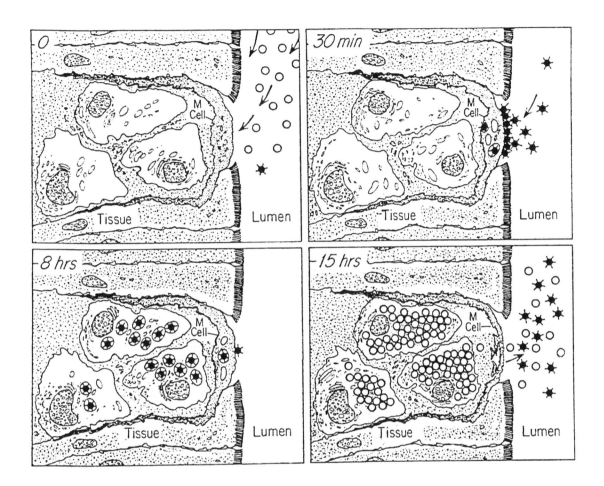

FIGURE 188

Reovirus replication in intestinal tissue after peroral inoculation. Virus is represented as open circles. ISVPs (intermediate size subviral particles) are represented as hexagons with protrusions representing an extended sigma 1 or cell attachment protein. (i) The intact virus of the inoculum (0-min panel) is digested to ISVPs upon entry into the intestinal lumen (30-min panel). (ii) Up to 8 hours post-inoculation (p.i.), virus continues to be present in the form of ISVPs (8-h panel). (iii) Between 8 and 15 min p.i., the viral titer increases by 10^3 PFU and replication has occurred so that the majority of particles are now intact virus (15-h panel). Some shedding of virus into the intestinal lumen results in digestion of replicated virus to ISPV (15-h-panel). Reovirus serotype 1 is shown penetrating the intestinal tissue via M cells overlying Peyer's patches (30-min panel), which subsequently are damaged as a result of reovirus replication (15-h panel). For simplification, a single M cell is shown with the virus replicating in undefined mononuclear cells beneath it. (From Bodkin, D.K., Nibert, M.L., and Fields, B.N., *J. Virol.*, 63, 4676, 1989. With permission.)

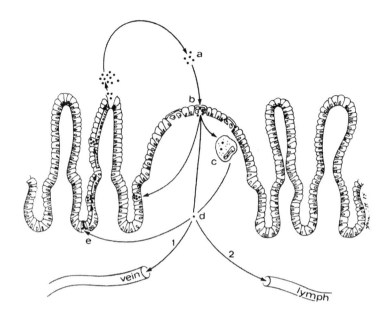

FIGURE 189

After oral inoculation, serotype 1 strain Lang was shown to specifically infect the epithelial cells of the ileum, while sparing the epithelial cells in the duodenum, jejunum, and colon. The initial site of replication was localized in cells of the crypts of Lieberkühn adjacent to Peyer's patches. Virus was subsequently found by immunoperoxidase staining in cells migrating up the crypt-villus complex through the ileum. The severity of the pathological changes in the ileum was proportional to the concentration of the viral inoculum. (From Rubin, D.H., Kornstein, M.J., and Anderson, A.O., *J. Virol.,* 53, 391, 1985. With permission.)

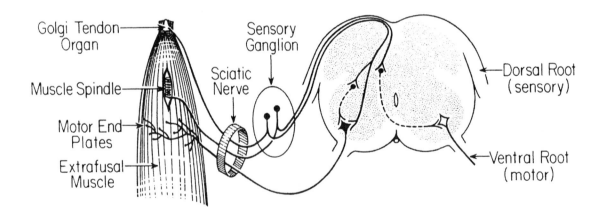

FIGURE 190

Potential sensory and motor neuronal pathways traveled by reovirus from hindlimb muscles to the lumbar spinal cord. Motor innervation of the muscle spindles as well as most connections of sensory and motor neurons have been omitted for clarity. Solid lines represent sensory and motor nerve fibers. Dashed lines represent interneurons. Categories of neurons which could be primarily or secondarily infected are indicated by filled cell bodies. (From Flamand, A., Gagner, J.-P., Morrison, L.A., and Fields, B.N., *J. Virol.,* 65, 123, 1991. With permission.)

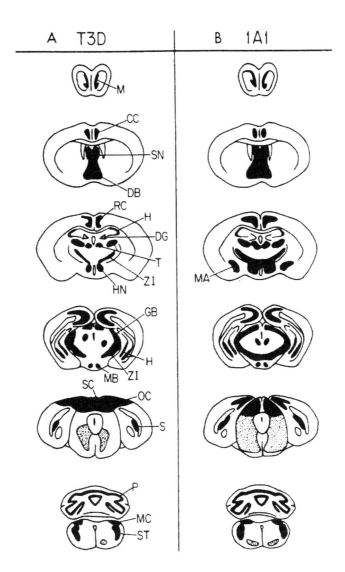

FIGURE 191

Coronal sections of brain in mice 12 days after i.c. inoculation with a lethal dose of virus. (A) Strain T3D-induced pathology; (B) Strain 1A1-induced pathology. Solid shading indicates areas of intense tissue destruction; stippled shading indicates areas of sparse pathologic changes. CC, cingulate cortex; DB, nucleus of the ventral limb of the diagonal band; DG, dentate gyrus; GB, lateral geniculate body; H, hippocampus; HN, hypothalamic nuclei; M, mitral cell layer; MA, medial amygdaloid nucleus; OC, occipital cortex; P, Purkinje cells; RC, retrosplenical cortex; S, subiculum; SC, superior colliculus, SN, septal nucleus; ST, spinal trigeminal nucleus; T, thalamic nuclei; ZI, zona inserta. (From Morrison, L.A., Fields, B.N., and Dermody, T.S., *J. Virol.*, 67, 3019, 1993. With permission.)

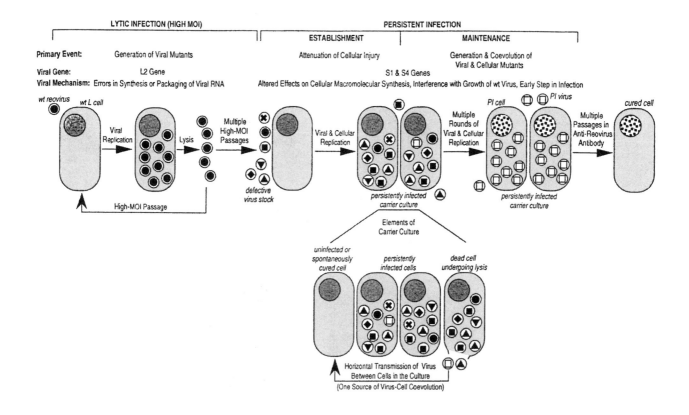

FIGURE 192

Model for persistent infection by reoviruses. Wild-type (wt) reoviruses are shown as solid circles. Mutant viruses arising or undergoing selection during serial passage at high MOI (multiplicity of infection) or during the maintenance of persistent infection (PI) are indicated by different symbols. Mutant cells undergoing selection during persistent infection are indicated by stippled nuclei. Viral genes *(top)* have been associated with different steps in the process by genetic analyses. (From Nibert, M.L., Schiff, L.A., and Fields, B.N., in *Fields Virology,* 3rd ed., Vol. 2, Fields, B.N., Knipe, D.M., and Howley, P.M., Eds.-in-chief, Lippincott-Raven, Philadelphia, 1996, 1557-1596. © Lippincott Williams & Wilkins. With permission.)

6.XVII. RETROVIRIDAE

Dimeric linear (+) sense ssRNA
Cubic, enveloped

This large family has seven genera, five of which are illustrated here. Viruses occur in mammals, birds, reptiles, and fishes. Particles are spherical, 80-100 nm in diameter, and consist of an envelope and an icosahedral or conical capsid which contains the genome and reverse transcriptase. Infecting viral RNA is reverse transcribed into linear dsDNA, which is then integrated into cellular DNA as a provirus. Integration is a prerequisite for replication. The provirus persists in the cell or is transcribed into novel viral RNA. Viruses are frequently associated with tumors (cancer, sarcomas, lymphomas), leukemias (alpha-, beta-, and gammaretroviruses), anemias and immunodeficiencies (lentiviruses), or disorders with symptoms of inflammation and autoimmunity (arthritis, encephalitis, mastitis, pneumonia). Transmission is exogenous or endogenous (via the germ line as integrated provirus). Exogenous transmission is horizontal (via blood, sexual contact, saliva, a.s.o.) or vertical (milk and placenta).

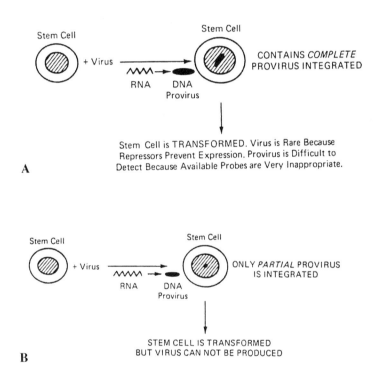

FIGURE 193

(A) Classical scheme of cell transformation by a RNA tumor virus. The infected cell here is a stem cell. Viral RNA is transcribed to DNA via reverse transcriptase. Newly synthetized provirus DNA is completely integrated into cellular chromosomal DNA. The provirus carries genes for leukemogenesis. When expressed, cells are transformed to leukemic cells. All their daughter cells contain at least one complete provirus. Virus is rarely seen because provirus is rarely *completely* expressed, but partial expression is sufficient for transformation. Provirus identification may be difficult because of inadequate probes (no homology between available animal viruses and human candidate viruses).

(B) Partial provirus model. As above, the stem cell is infected by a RNA tumor virus and transformed to a leukemic cell. However, only a *partial* provirus is synthetized and/or integrated into the host cell chromosomal DNA. Daughter cells will be transformed and contain partial provirus. The latter carries sufficient information to transform the cell, but a virus cannot usually be isolated. Viral proteins are limited and difficult to detect. Provirus would be very difficult to detect. (Author's note: the same chapter proposes three other models.) (Adapted from Gallo, R.C., in *Viruses and Environment,* Kurstak, E. and Maramorosch, K., Eds., Academic Press, New York, 1978, 43. With permission.)

A. GENUS *ALPHARETROVIRUS*

Member of this genus, formerly designated as "avian type C" retroviruses, infect birds and include avian leukosis virus (ALV) and the famous Rous sarcoma virus. Viruses are able to induce many types of disease; for example, ALV causes not only leukemia, but also erythroblastosis, anemia, and, more rarely, carcinomas and sarcomas. ALV infection of chickens is very common. The virus is transmitted horizontally, congenitally, or through the germ line.

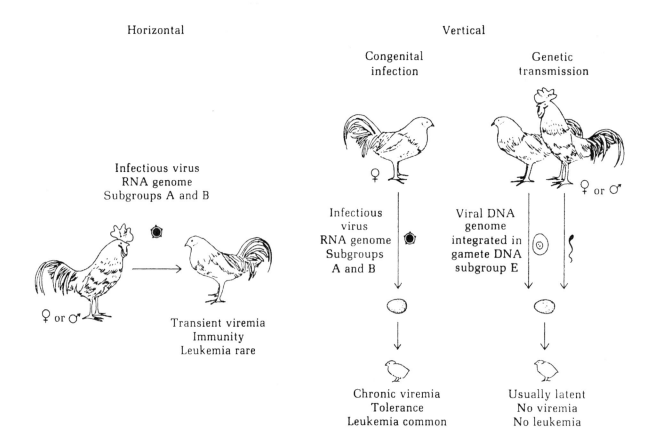

FIGURE 194

Transmission of avian leukosis (leukemia) virus. Exogenous viruses are transmitted horizontally and congenitally. Endogenous virus transmission is via the germ line. Viruses proliferate in B cells. In rare cases the provirus integrates near the *myc* and B-cell lymphomas are produced. (From Volk, W.A., *Essentials in Microbiology*, J.B. Lippincott, Philadelphia, 1978, 572. © Lippincott Williams & Wilkins. With permission.)

B. GENUS *BETARETROVIRUS*

Type species and only member is the mouse mammary tumor virus, characterized by an eccentric core and formerly known as type B retrovirus. Transmission is mostly by milk or endogenously as provirus. The virus causes mammary carcinomas and lymphomas.

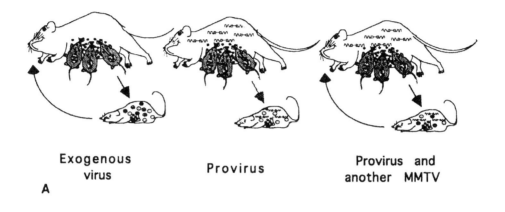

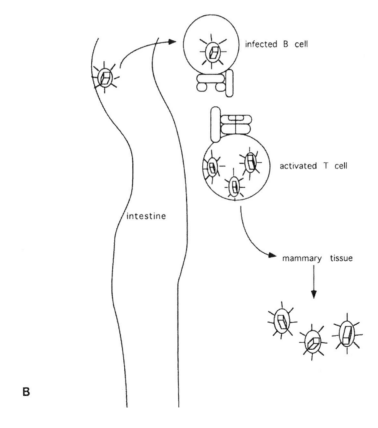

FIGURE 195

(A) Transmission of infectious mouse mammary tumor virus (MMTV) in the presence or absence of proviral MMTV DNA in the germ line. (B) MMTV in the gut infects B lymphocytes, which leads to superantigen expression. T cells are activated and replicate the virus. (From Huber, B.T., Hsu, P.N., and Sutkowski, N., *Microbiol. Rev.*, 60, 473, 1996. With permission.)

C. GENUS *GAMMARETROVIRUS*

This genus corresponds to the former "mammalian type C and D" retroviruses. Members include the Friend and Moloney murine leukemia viruses (MuLVs) and are also found in birds and reptiles. MuLVs cause malignant tumors (leukemias, sarcomas, lymphomas) and also autoimmune disorders similar to human lupus erythematosus. Viruses are exogenous or endogenous. Feline leukemia has been studied in particular detail. The disease is self-limiting in most cases, but may become persistent and evolve into leukemia (or anemia or lymphosarcomas) and cause an immunodeficiency syndrome.

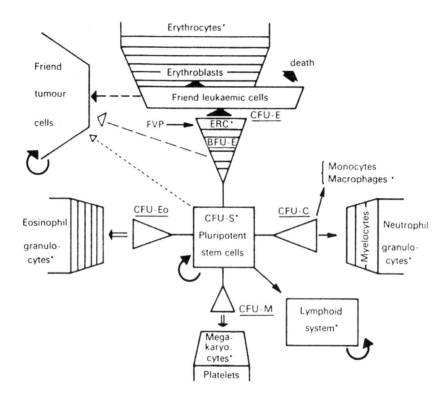

FIGURE 196

Leukemogenesis and tumorigenesis induced by Friend leukemia virus. Oncogenic potency (induction of excessive erythroid proliferation) is expressed only at the late erythroid precursor cell level, probably at the point when normal erythroid cells require erythropoietin responsiveness. After viral transformation, which takes about 30 hours, leukemic hyperbasophilic Friend cells multiply and differentiate along the erythrocyte pathway. Erythropoietin is not required for pathological differentiation. As indicated by kinetic studies, Friend leukemic cells are not self-maintaining (they die or differentiate). Disease progression requires a constant recruitment of target cells from the ERC and CFU compartments. Viral replicative ability is not restricted to erythroblastic cells. Hematopoietic stem cells and the various hematopoietic lineages are also infected (*). Their development remains qualitatively normal but is quantitatively affected. Later in the disease the tumor cell population arises (broken line) from a yet unidentified target cell for tumorigenesis. These tumor cells have acquired an infinite lifespan. Cell populations marked with an *asterisk* are infected by at least one component of the Friend virus complex. BFU, burst-forming unit; CFU, colony-forming unit; C, neutrophilic granulocyte and macrophage; E, erythroid; Eo, eosinophilic granulocyte; ERC, erythropoietin-responsive cell; FVP, Friend virus particle; M, megakaryocytic; S, self-maintaining population with a large majority of cells in G_0. (From Tambourin, P.E., in *Stem Cells and Tissue Homeostasis*, 2nd Brit. Symp. Cell Biol., Lord, B.I., Potten, C.S., and Cole, R.J., Eds., Cambridge University Press, Cambridge, 1978, 259. Reprinted with the permission of Cambridge University Press.)

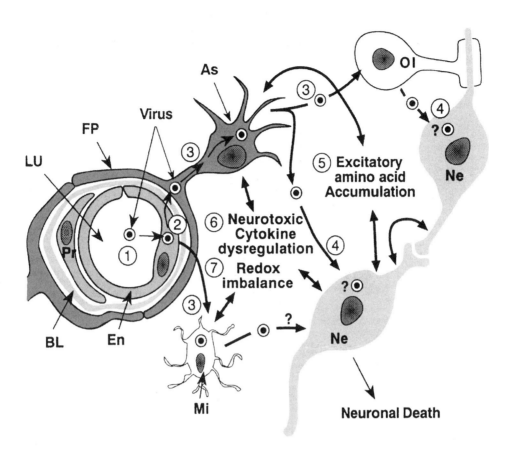

FIGURE 197

Possible mechanisms of MuLV-induced neuronal damage. (1) After early replication in peripheral sites, the virus spreads via the circulatory system to endothelial cells of the CNS. (2) From there, the virus passes through the blood-brain barrier into the CNS. (3) Within the CNS parenchyma, the virus infects microglial cells, astrocytes, and oligodendrocytes. (4) Neurons are either not infected or, if infected, do not express readily detectable viral antigens, yet display the most obvious cytopathic effect. (5) The virus may participate indirectly in the killing of neurons by disruption of microglial cells, astrocytes, and oligodendrocytes, which are required for maintenance and viability of the surrounding neurons. For example, astrocytes may fail to maintain normal glutamate levels in the extracellular space. Accumulation of this excitatory amino acid may be neurotoxic. (6) Virus-infected astrocytes and microglial cells may promote overproduction of certain cytokines such as IL-1, IL-6, and TNF which may damage the nearby neurons. (7) Virus infection of astrocytes and microglial cells may cause failure of redox homeostasis modulated by these cells with subsequent death of nearby neurons. As, astrocyte; BL, basal lamina: En, endothelial cell; FP, foot process; LU, lumen; Mi, microglial cell; Ne, neuron; Ol, oligodendrocyte. (From González-Scarano, F., Nathanson, N., and Wong, P.K.Y., in *The Retroviridae*, Vol. 4, Levy, J.A., Ed., Plenum Press, New York, 1992, 409. With permission.)

FELINE LEUKEMIA

Feline leukemia is characterized by persistent viremia, lymphopenia, progressive weight loss, enteropathy, and opportunistic infections. There are three stages: (1) early host-virus interaction: (2) persistent infection with immunosuppression in susceptible cats vs. self-limiting infection and immunity in resistent cats, and (3) leukemia and related diseases or fatal opportunistic infection.

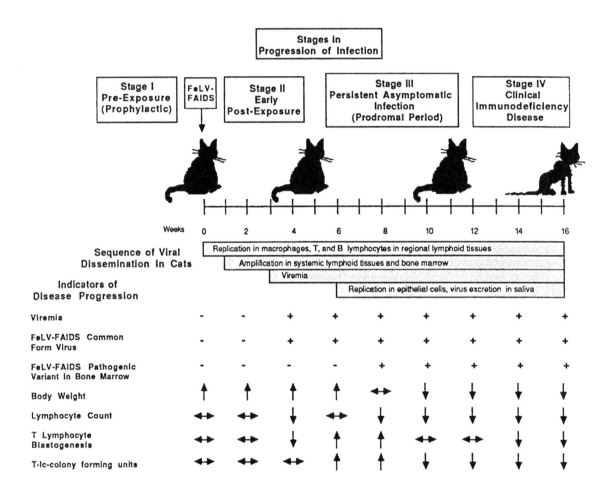

FIGURE 198

Events in FeLV-FAIDS-induced immunodeficiency syndrome. T-lc, T lymphocyte. (Author's note: FeLV-FAIDS designates a particular strain of feline leukemia virus that constantly induces a fatal immunodeficiency syndrome in inoculated cats and not a lentivirus.) (From Hoover, E.A., Zeidner, N.S., Perigo, N.A., Quackenbush, N.L., Strobel, J.D., Hill, D.L., and Mullins, J.I., *Intervirology,* 30 (suppl. 1), 12, 1989. With permission.)

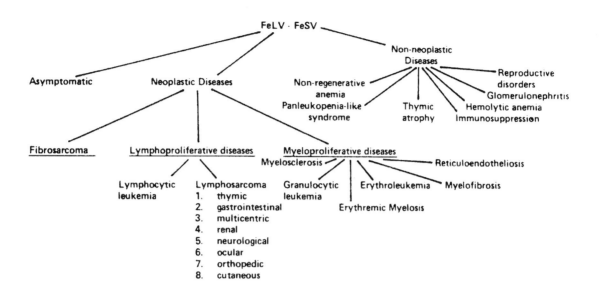

FIGURE 199

Scope of clinical diseases produced or related to the feline leukemia virus and feline sarcoma virus. (Reprinted with permission from Hause, W.R. and Olsen, R.G., in *Feline Leukemia,* Olsen R.G., Ed., CRC Press, Boca Raton, 1981, 89. Copyright CRC Press, Boca Raton, Florida.)

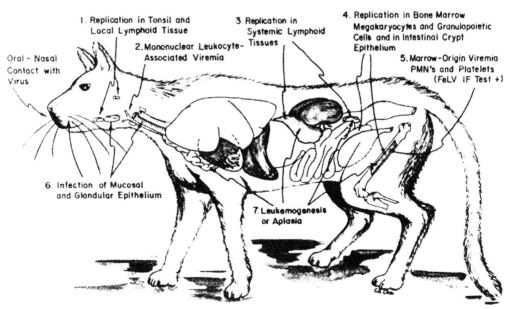

FIGURE 200

Sequence of events in the pathogenesis of feline leukemia virus infection. (Reprinted with permission from Hoover, E.A., Rojko, J.L., and Olsen, R.G., in *Feline Leukemia,* Olsen, R.G., Ed., CRC Press, Boca Raton, 1981, 31. Copyright CRC Press, Boca Raton, Florida.)

D. GENUS *DELTARETROVIRUS*

Viruses of this small genus are found in humans, monkeys, and cattle. One member, human T-cell leukemia virus type 1 (HTLV-1) was the first human retrovirus to be discovered. The virus causes T cell leukemia and myelopathy with spastic paresis in adults. Cases cluster in southwest Japan. HTLV-1 pathogeny seems to be linked to the viral Tax protein. Transmission is horizontal. Cattle viruses are associated with B cell lymphoma.

FIGURE 201

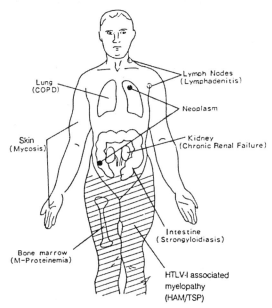

HTLV-1 and secondary immunodeficiency: clinical features of HTLV-infection. The predominant physical features are peripheral lymph node enlargement, hepato- and splenomegaly, and skin lesions. Hyper-calcemia is frequent. Causes of death are pulmonary complications, including *Pneumocystis carinii* infection, hypercalcemia, *Cryptococcus* meningitis, and disseminated herpes zoster, and intravascular coagulopathies. (Author's legend) (From Yamaguchi, K., Kiyokawa, T., Futami, G., Ishii, T., and Takatsuki, K., in *Human Retrovirology: HTLV*, Blattner, W.A., Ed., Raven Press, New York, 1990, 163-171. © Lippincott Williams & Wilkins. With permission.)

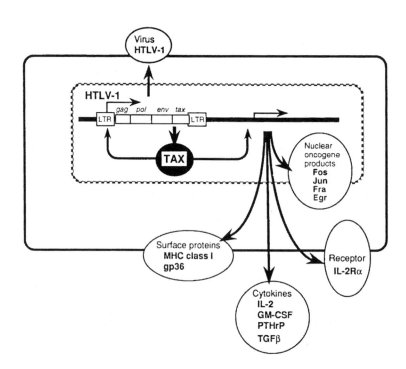

FIGURE 202

The viral nuclear protein Tax *trans*-activates the viral LTR (long terminal repeat) to enhance expression of viral genes, and also *trans*-activates cellular genes, some of which probably modulate cellular phenotypes such as immortalization and transformation. (Reprinted from *Trends Microbiol.*, 1, 131-135, 1993. Yoshida, M., HTLV-1 Tax: regulation of gene expression and disease. © 1993, with permission from Elsevier Science.)

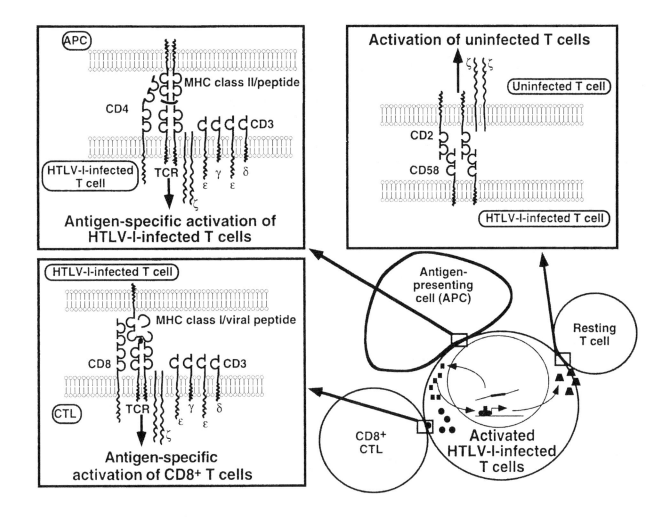

FIGURE 203

Activation of T cells by HTLV-I. Infection of CD4$^+$ T cells influences immune system T-cell activation by at least four separate pathways. (i) The HTLV-I-infected T cells are activated by viral interference with signaling pathways and transcriptional regulation (bottom right). (ii) The infected T cell interacts with and activates resting T cells (top right) in a viral antigen-independent manner. The CD58-CD2 interaction is critical, but other molecular interactions and cytokines (not shown) are likely to contribute. (iii) Virus-specific CD8$^+$ cells (and to a lesser degree CD4$^+$ cells) are activated by recognition of viral peptide epitopes (bottom left). (iv) APC may present MHC class II-restricted peptide antigens that activate the HTLV-I-infected cell (top left). This activation process is altered by virtue of viral interference with the signaling cascade or the transcriptional regulation of the HTLV-I-infected T cell, or both. (From Höllsberg, P., *Microbiol. Mol. Biol. Rev.*, 63, 308, 1999. With permission.)

E. GENUS *LENTIVIRUS*

Particles are characterized by a cone- or bar-shaped capsid and occur in primates, ungulates (cattle, horses, sheep, goats), and cats. Viruses are horizontally transmitted and very host-specific. They are neurotropic rather than oncogenic and infect cells of the immune system, particularly monocyte/macrophages. Viruses frequently affect the hematopoietic and central nervous systems, but also cause such diverse diseases as arthritis-encephalitis (goats), pneumo-encephalitis (Maedi-Visna) of sheep, and autoimmune lesions. Diseases are characterized by long incubation periods and persistence of viruses in cells of the macrophage lineage. Lentiviruses can be divided into two subgroups, those that induce immunodeficiency and those that do not. Immunodeficiency viruses (human, simian, feline) cause a variety of disorders that result from virus multiplication and from opportunistic infections due to the immunocompromised status of the host.

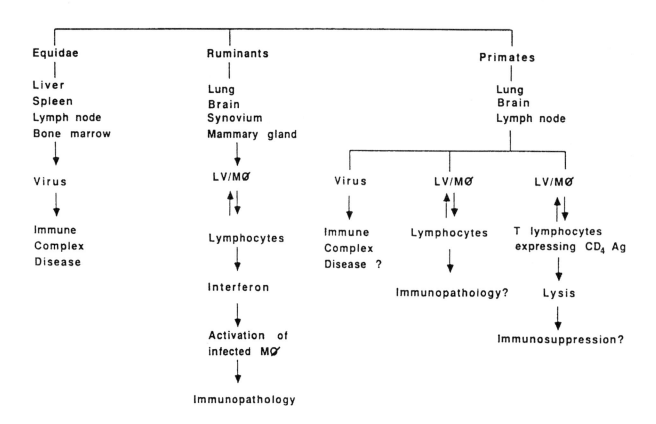

FIGURE 204

Fate of lentiviruses (LV) produced by macrophages (Mφ) in different host animals. Free viruses produced in equine macrophages induce and combine with antibodies to produce pathologic immune complexes. Ruminant animal macrophages are poorly permissive for virus replication, but these cells present lentiviral antigens to lymphocytes. Various cytokines, including interferons, are produced. These act back on infected macrophages, causing enhancement in expression of Ia antigens. This leads to immunopathology. Primate macrophages vary in production of lentiviral antigens, and responses resembling those in both equidae and ruminants are obtained. In addition, the high affinity of the viral glycoprotein for the CD4 site on primate T4 lymphocytes results in fusion between the antigen-presenting macrophage and the T4 cell. This results in lysis of the latter and gradual elimination of this subset of lymphocytes, leading to the well-known immunosuppression of these host animals. (From Narayan, O. and Clements, J.E., in *Virology,* 2nd ed., Vol. 2, Fields, B.N. and Knipe, D.M., Eds.-in-chief, Raven Press, New York, 1990, 1571-1589. © Lippincott Williams & Wilkins. With permission.)

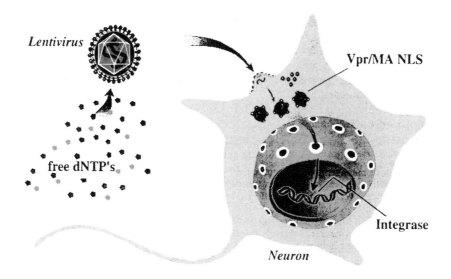

FIGURE 205

Lentivirus nuclear import mechanism. The recognition of the uncoated lentivirus nucleoprotein complex, here the mutant lentivirus vector Vpr/MA NLS, by the cell import machinery allows the active transport of viral genome through the nucleopore and stable integration into the target cell genome. This mechanism allows lentiviruses to infect nondividing cells. Reverse transcription is promoted and there is a subsequent increase in infection efficiency following the addition of the four dNTPs in vitro. (From Blömer, U., Naldini, L., Kafri, T., Trono, D., Verma, I.M., and Gage, F.H., *J. Virol.*, 71, 6641, 1997. With permission.)

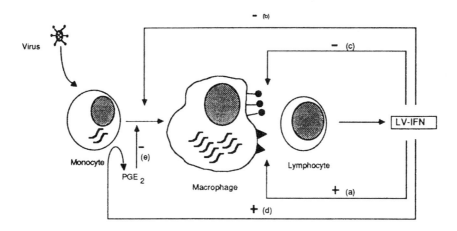

FIGURE 206

Lentivirus IFN inhibits the proliferation and maturation of monocytes and, indirectly, the replication of caprine arthritis-encephalitis virus (CAEV) in goats. Lentivirus-infected monocytes become more permissive for viral replication as they mature to macrophages. The latter produce abundant viral RNA (bent filaments) and express viral antigen (pinheads), partly by way of their class II MHC antigens (triangles). This is a prerequisite for subsequent production of LV-IFN by lymphocytes. The various effects of LV-IFN are interdependent. LV-IFN enhances the expression of class II MHC antigens on the macrophage (a), which stimulates the further production of LV-IFN. At the same time, it inhibits the viral replicative cycle, both by inhibiting monocyte proliferation and maturation (b) and by blocking transcription (c). The resulting reduction of viral protein expression downregulates monocyte-macrophages to produce PGE_2 (d), which itself has an inhibitory effect on monocyte proliferation (e). (From Zink, M.C. and Narayan, O., *J. Virol.*, 63, 2578, 1989. With permission.)

HUMAN ACQUIRED IMMUNODEFICIENCY SYNDROME (AIDS)

The disease is caused by human immunodeficiency virus (HIV) types 1 and 2, both closely related to several simian immunodeficiency viruses (SIVs). The virus is essentially transmitted through sexual contact and blood and persists as an integrated provirus in lymphoid organs that serve as HIV reservoirs and primary sites for virus replication. The virus persists despite potent cytotoxic T lymphocyte (CTL) responses. Viral persistence leads to chronic stimulation of the immune system and destruction of lymphoid tissues.

The initial acute illness is influenza-like with fever, chills, adenopathy, viremia, and sometimes neurological symptoms. Viral levels then fall and the infected organism enters a latent period of variable length (several months to over 10 years). Subsequently, HIV causes an insidious, progressive deterioration of immune functions, characterized by depletion of CD4$^+$ T lymphocytes, that leads to profound immunosuppression and the emergence of opportunistic infections (for example by *Candida albicans* or HSV) and neoplasms (Kaposi's sarcoma, lymphomas). HIV infection also leads to neurologic complications ranging from meningitis to dementia. The typical course of HIV infection, always fatal, involves a period of 8 to 10 years.

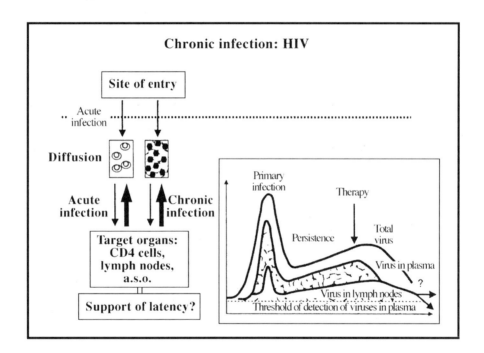

FIGURE 207

Human immunodeficiency (HIV): a model of acute infection evolving toward chronicity. HIV persists by replicating continually, mainly in mononuclear blood cells and certain cells of lymph nodes. The evolution of the chronic infection can be monitored, notably at the onset of therapy, by quantifying infected circulating cells or viral constituents in the plasma (RNA, p24 antigen). In association with other investigative techniques (e.g., of virus in lymph nodes), the available markers permit some evaluation of virus movement between its various biological reservoirs. (From Maréchal, V., Dehée, A., and Nicolas, J.-C., *Virologie,* 1, special issue, 11, 1997. With permission of John Libbey Eurotext.)

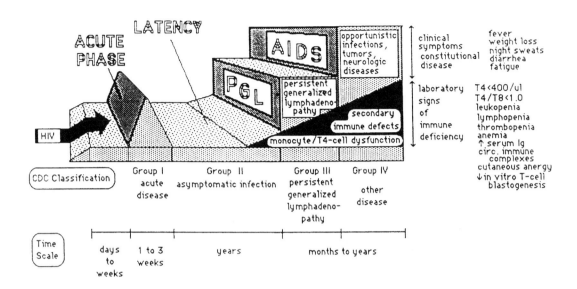

FIGURE 208

Stages of HIV infection. The infection may be followed by an acute mononucleosis-like syndrome. After a latency period of variable length, the infection leads to progressive deterioration of immune functions. The primary immune defect in AIDS consists of the incapability of T4-cells to recognize antigen presented to them by monocytes/macrophages. Secondary defects result from dysfunction of cells that depend on T4-cell help. The immune defect results in persistent generalized lymphadenopathy (PGL) and later in disease with constitutional symptoms, opportunistic infections, tumors, and neurological disease (AIDS). (Fig. 10 from Schüpbach, J., in *Curr. Topics Microbiol. Immunol.,* 142, 36, 1989. © Springer-Verlag. With permission.)

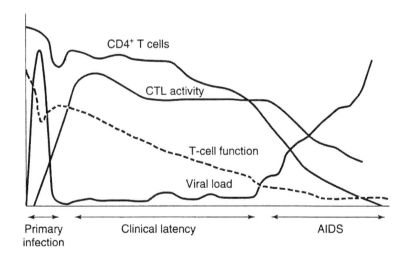

FIGURE 209

The course of HIV infection. After the initial period of primary infection (several weeks) viral load decreases and a period of clinical latency starts, which can last from <1 year to >10 years. In general, T-cell functions decline and CD4+ T-cell counts decrease gradually during this period. Prior to the onset of AIDS, there is an increase in viral load and a loss of cytotoxic T lymphocyte (CTL) function. With progressive loss of CD4+ T-cells, symptoms start to occur and AIDS is diagnosed. From diagnosis, it usually takes several years before death. (Reprinted from *Trends Microbiol.,* 6, 144-147, 1998. Wolthers, K.C. and Miedema, F., Telomeres and HIV-1 infection: in search of exhaustion. © 1998, with permission of Elsevier Science.)

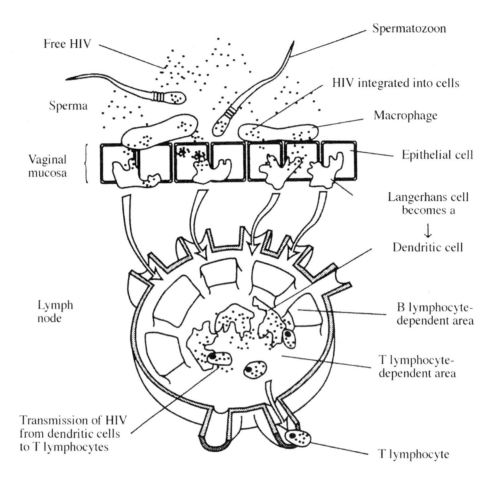

FIGURE 210

Sexual transmission of HIV. (From Bergeron, M.-G., in *VIH/Sida. Une approche multidisciplinaire*, Reidy, M. and Taggart, M.-E., Eds., Gaëtan Morin, Montréal, 1995, 567. With permission.)

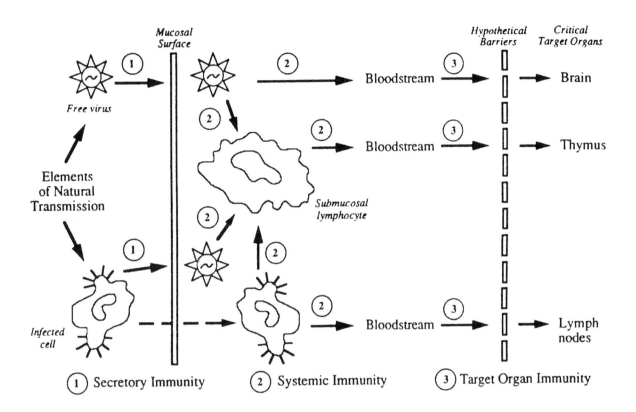

FIGURE 211

Mucosal and systemic immunity to HIV infection. (1) Secretory immunity is expected to play its major role in reducing transmission of virus and/or infected cells across mucosal surfaces. (2) Breakthrough infections would then be met by systemic immunity, composed of neutralizing antibodies, cytotoxic lymphocytes, and antibody-armed killer cells (ADCC). (3) There may also be barriers that are important for transmission of the infection from the bloodstream to critical target organs. The blood-brain barrier is an example but it is less clear whether there are any limiting factors for infection of the thymus or lymph nodes. (From Bolognesi, D.P., *Adv. Virus Res.*, 42, 103, 1993. With permission.)

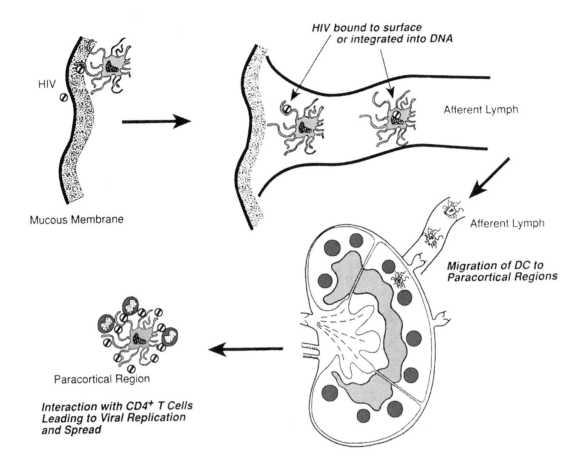

FIGURE 212

Role of dendritic cells (DC) in the initiation of viral replication. HIV enters a mucous membrane (sexual spread) or skin (needle injury) and binds to or infects a tissue DC. Upon receiving the appropriate signal, the DC travels in the afferent lymphatics, enters a draining lymph node, and migrates through the subcapsular sinus to the paracortical region. Within the paracortical region, the DC interacts with and activates CD4[+] T cells, leading to productive infection and spread of HIV. (From Weissman, D. and Fauci, A.S., *Clin. Microbiol. Rev.,* 10, 358, 1997. With permission.)

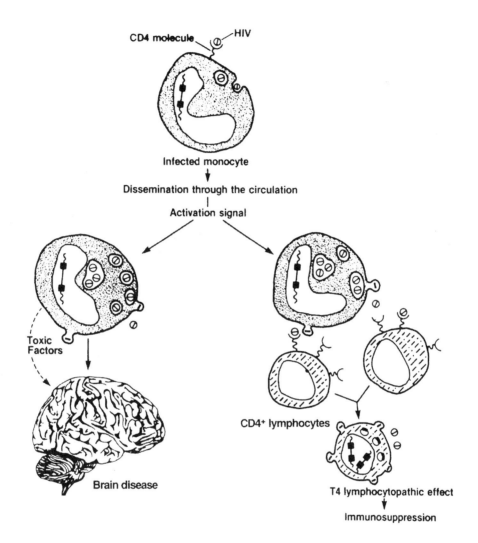

FIGURE 213

The role of the monocyte/macrophage in HIV infection. The HIV-infected monocyte/macrophage presents virus to T4 lymphocytes which are subsequently infected and killed. In addition, infected monocyte/macrophages transport HIV to the brain where they can either infect CD4-bearing cells in the brain or produce neurotoxic factors. (From Fauci, A.S., *Trans. Assoc. Am. Physicians*, 101, clx, 1988. Reprinted by permission of Blackwell Science, Inc.)

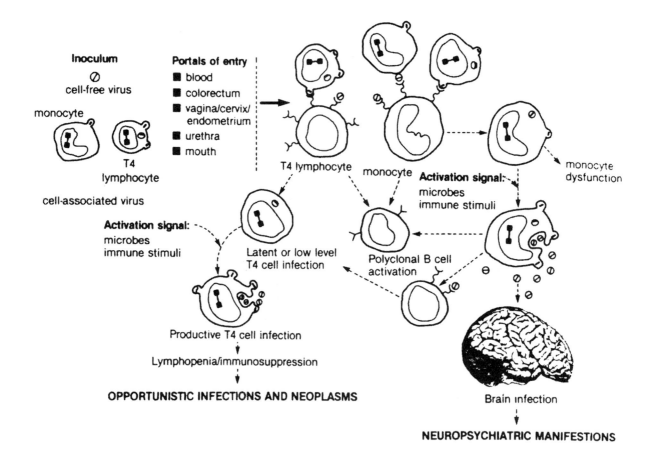

Inoculum
⊘
cell-free virus

monocyte

T4
lymphocyte

cell-associated virus

Portals of entry
- blood
- colorectum
- vagina/cervix/
  endometrium
- urethra
- mouth

T4 lymphocyte   monocyte

monocyte
dysfunction

Activation signal:
microbes
immune stimuli

Activation signal:
microbes
immune stimuli

Latent or low level
T4 cell infection

Polyclonal B cell
activation

Productive T4 cell infection

Lymphopenia/immunosuppression

OPPORTUNISTIC INFECTIONS AND NEOPLASMS

Brain infection

NEUROPSYCHIATRIC MANIFESTIONS

FIGURE 214

Potential mechanisms of pathogenesis of HIV infection. Virus enters the body through a variety of portals of entry either as cell-free virions or in a cell-associated form. The virus binds to and infects CD4+ cells and monocytes. The monocytes may also phagocytize virus. HIV infection can exist either as a latent or a low-level or chronic form. Upon activation of the infected cell, virus is produced resulting in a cytopathic effect on T cells and to a much less extent on monocytes. The monocyte can serve as a reservoir for HIV, transporting the virus to various parts of the body, particularly the brain, thereby leading to neuropsychiatric abnormalities. In addition, monocytes can pass the virus to T4 cells. Cytopathicity follows upon activation of the T4 cells, with resulting cell death and immunosuppression leading to the development of opportunistic infections. (Reprinted from *Science*. From Fauci, A.S., *Science,* 239, 617,1988.)

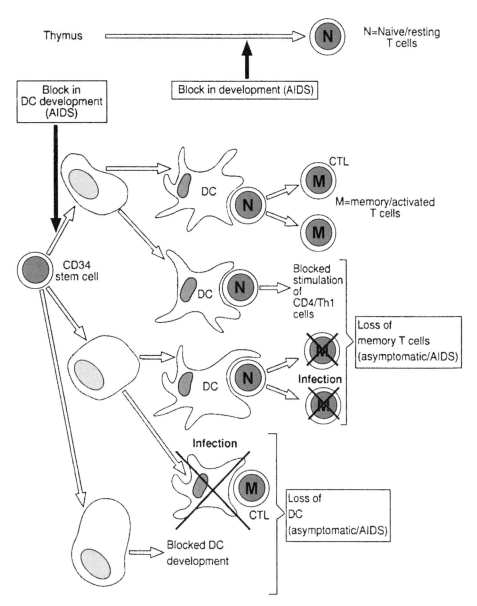

Consequences of HIV infection on DC development and function

FIGURE 215

Development and cellular interactions of dendritic cells (DC) and naive (N) or memory (M) T cells in the production of immunodeficiency during HIV infection. Boxed texts show possible major defects. Crosses denote destruction of cells by direct infection, syncytial formation, or killing by cytotoxic T lymphocytes (CTL). (From Knight, S.C. and Patterson, S., *Annu. Rev Immunol.*, 15, 593, 1997. With permission from the *Annual Review of Immunology*, © 1997, by Annual Reviews http://www.AnnualReviews.org.)

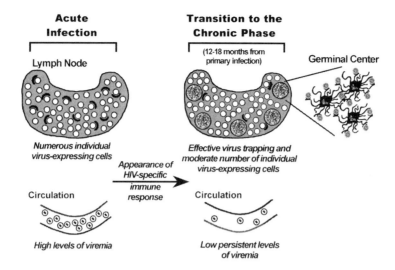

FIGURE 216

Virologic and histopathologic events occurring in lymphoid organs and circulation during acute HIV infection and during the transition to the chronic phase. After entry, HIV localizes in the lymphoid organs that represent the primary anatomic site for the establishment of HIV infection. Early in acute infection, numerous individual virus-expressing cells are detected in lymph nodes, and the peak of virus-expressing cells precedes and/or coincides with that of viremia. Furthermore, early in acute infection, germinal center formation is minimal or absent. After transition to the chronic phase, the number of virus-expressing cells declines significantly. Simultaneously with extensive germinal centers formation, virus particles are efficiently trapped within the FDC (follicular dendritic cell) network, and viremia is dramatically downregulated. (From Pantaleo, G. and Fauci, A.S., *Annu. Rev. Microbiol.,* 50, 825, 1996.)

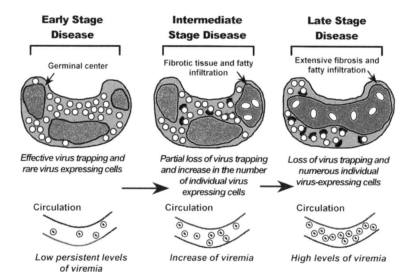

FIGURE 217

Changes in virus distribution, histopathology, and viremia associated with the progression of HIV disease from early to late stages. In early-stage disease, lymph nodes present extensive germinal center formation (germinal centers of large size and irregular shape), effective virus trapping in the FDC network *(grey area)* , very rare virus-expressing cells, and low levels of viremia. In intermediate-stage disease, large areas of the lymphoid tissue undergo involution (i.e., fibrosis and fatty infiltration), and both the number of virus-expressing cells in lymph nodes and the levels of viremia increase. In late-stage disease, most of lymphoid tissue is replaced by fibrosis and fatty infiltration, virus trapping is generally lost, and there are numerous virus-expressing cells and high levels of viremia. (From Pantaleo, G. and Fauci, A.S., *Annu. Rev. Microbiol.,* 50, 825, 1996.)

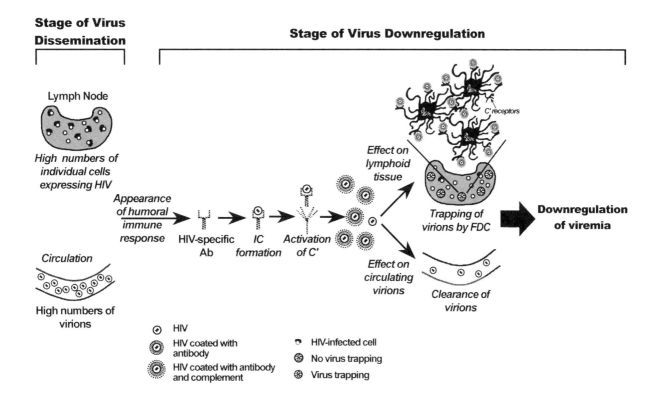

FIGURE 218

Downregulation of viremia by HIV-specific humoral immune response. A large number of antibodies against a variety of HIV proteins are produced, and a fraction of these antibodies may bind C'. These C'-binding antibodies may form immune complexes (IC) with virus particles, and these IC may be trapped within the FDC network (FDC express on the cell surface receptors for C') and the reticuloendothelial system. These events may significantly contribute to the clearance of virions from circulation. (From Pantaleo, G. and Fauci, A.S., *Annu. Rev. Microbiol.,* 50, 825, 1996.)

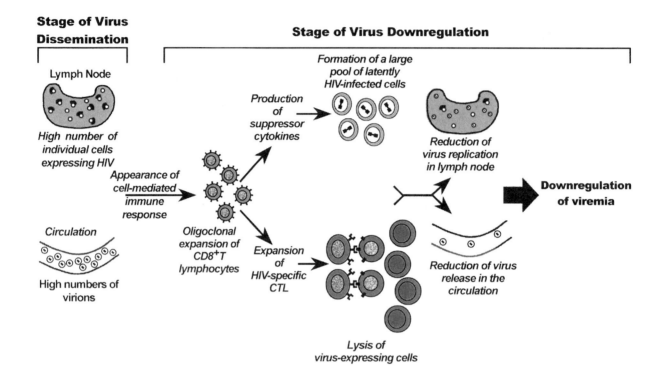

FIGURE 219

Downregulation of viremia by HIV-specific cell-mediated immune response. Following virus dissemination and generally in coincidence with peak viremia, oligoclonal expansions of CD8$^+$ T lymphocytes represent the major component of the HIV-specific cell-mediated immune response. These CD8$^+$ lymphocytes mediate either cytotoxic activity or produce a number of soluble factors that may suppress virus replication. Therefore, HIV-specific cell-mediated immune response may contribute to, or play the prime role in, the downregulation of viremia in primary infection, both by lysis of virus-expressing cells and by suppression of virus replication. (From Pantaleo, G. and Fauci, A.S., *Annu. Rev. Microbiol.,* 50, 825, 1996.)

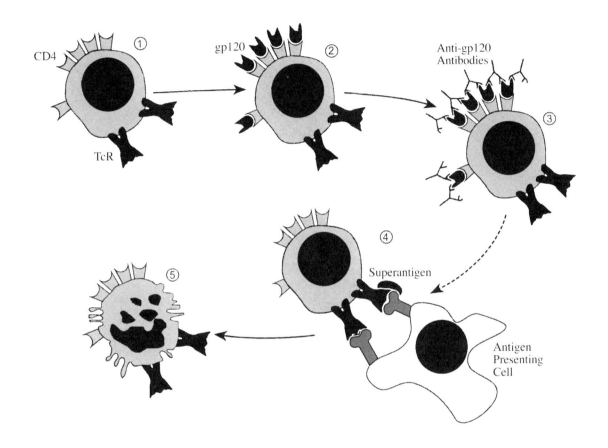

FIGURE 220

Hypothetical mechanism for depletion of uninfected CD4+ cells in HIV-1-infected individuals by induction of apoptosis. T-helper cells (1) bearing CD4 and T-cell receptors (TcR) become coated with free gp120 (2) via specific interaction with the CD4 receptors. These gp120 molecules are then cross-linked by gp120-specific antibody (3), priming the cell for apoptosis. Later, a second signal resulting from antigen-specific recognition or superantigen activation (4) triggers the cell to undergo apoptosis. (From Norley, S. and Kurth, R., in *The Retroviridae,* Vol. 3, Levy, J.A., Eds. Plenum Press, New York, 1992, 363. With permission.)

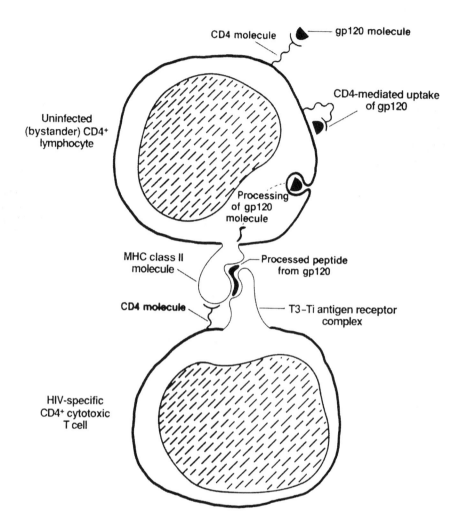

FIGURE 221

Elimination of uninfected CD4$^+$ T cells exposed to soluble gp120 by HIV-specific CD4$^+$ cytotoxic T cells. (From Rosenberg, Z.F. and Fauci, A.S., *Adv. Immunol.*, 47, 377, 1989. With permission.)

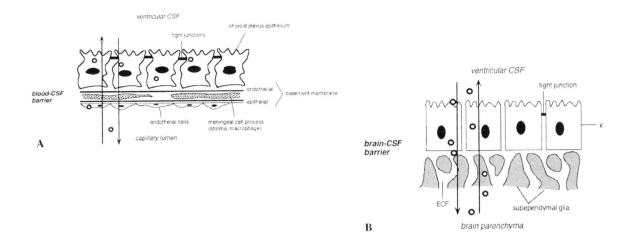

FIGURE 222

(A) Blood-CSF barrier to HIV-1 passage. An additional means of virus passage from the brain parenchyma to the CSF (and reverse) involves endothelial cell infection and release of cells into the stroma of the choroid plexus. Here infection of stromal cells and macrophages may occur, followed by infection of choroid plexus epithelium. (B) Brain-CSF barrier to HIV-1 passage. Transition of virus from the brain parenchyma into the ventricular CSF and vice versa involves infection of ependymal cells and subependymal glia, although there is no experimental evidence for ependymal cell infection. ECF, extracellular fluid. (Adapted from Kolson, D.L., Lavi, E., and González-Scarano, F., *Adv. Virus Res.,* 50, 1, 1998. With permission.)

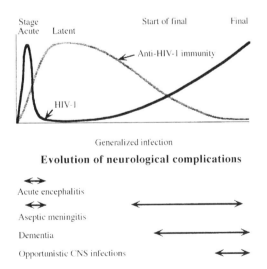

FIGURE 223

Evolution of neurologic AIDS complications, especially of dementia, in relation to the stage of HIV-1 infection. Acute encephalitis and meningitis are observed in the early stage; chronic meningitis and dementia may be seen in later and final stages. (From Piot, P., Kapita, B.M., Ngugi, E.N., Mann, J.M., Colebunders, R., and Wabitsch, R., *Le SIDA en Afrique, Manuel du Praticien*, World Health Organization, Geneva, 1993, 36. With permission.)

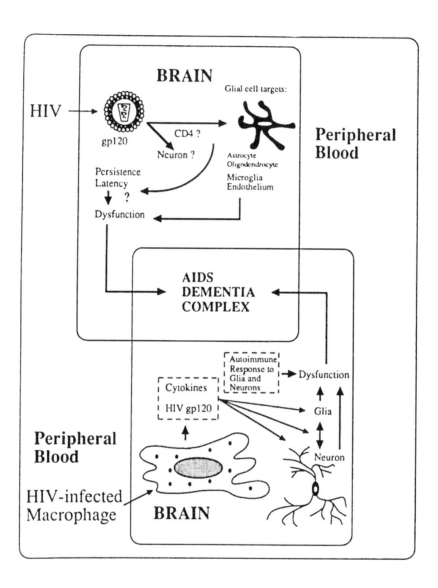

FIGURE 224

Mechanisms of neurologic dysfunction associated with HIV infection. Damage to the nervous system as a result of HIV infection may occur via two general pathways. One mechanism involves direct virus infection of nervous cell system cell types that may result in either killing or alteration of cellular metabolic function. The retention of the viral genome in neural cells in a persistent or latent form may also lead to neuronal or glial cell dysfunction. A second general pathway centers on the cytotoxic action of HIV-specific products or aberrantly secreted cellular products that may interfere with nervous system function. In addition, autoimmune mechanisms and such factors as virus strain may also play a role in nervous system dysfunction. (Adapted from Wigdahl, B. and Kunsch, C., *Prog. Med. Virol.*, 37, 1, 1990. With permission.)

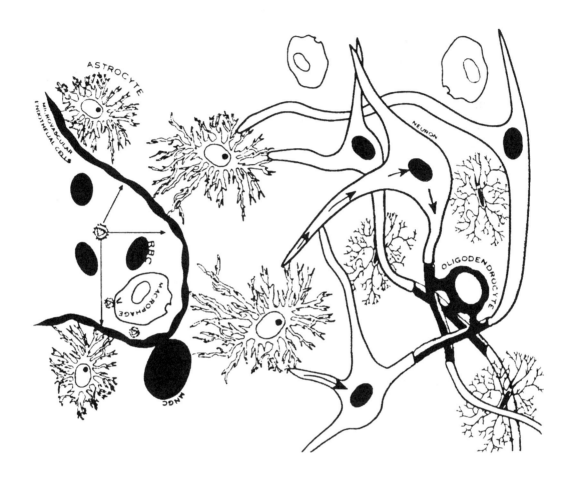

FIGURE 225

Potential mechanisms of early HIV-1 transmission to the central nervous system. Microvascular endothelial cells (MVEC) are infected either by cell-free virions or by HIV-1-infected mononuclear cells. MVEC then transfer HIV-1 to the other side of the blood-brain barrier by productive infection. Multinucleate giant cell (MNGC) formation results from infection of microglia and macrophages, subsequently spreading the viral infection either directly to neurons or individually via infected astrocytes and oligodendrocytes. Later, infection can seriously hamper the myelination of neurons, indirectly affecting neuronal functions. RBC, red blood cell. (From Bagasra, O., Lavi, E., Bobroski, L., Khalili, K., Pestaner, J.P., Tawadros, R., and Pomerantz, R.J., Cellular reservoirs of HIV-1 in the central nervous system of infected individuals: identification by the polymerase chain reaction and immunohistochemistry, *AIDS*, 10, 573-585, 1996. © Lippincott Williams & Wilkins. With permission.)

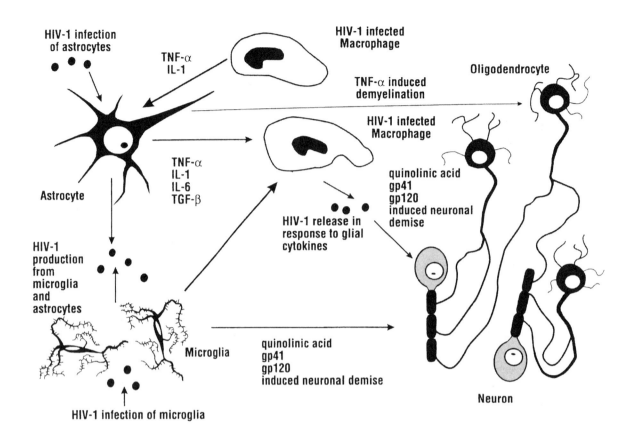

FIGURE 226

Proposed roles of cytokines in the pathogenesis of HIV-1 infection of the nervous system. HIV-1 most likely enters the CNS in an infected macrophage. Microglial cells and astrocytes may then become infected with HIV-1. TNF-α, IL-1, and IL-6 produced by microglia and astrocytes may amplify HIV-1 multiplication in macrophages. TGF-β secreted by astrocytes may recruit additional macrophages into the region. Microglia and macrophages release toxic viral components and quinolinic acid that can damage neurons. This cycle of immune activation within brain parenchyma most likely leads to an increased viral burden in the brain as well as to the broad spectrum of neurological disease caused by HIV-1 infection in the brain. (From Atwood, W.J., Berger, J.R., Kaderman, R., Tornatore, C.S., and Major, E.O., *Clin. Microbiol. Rev.*, 6, 339, 1993. With permission.)

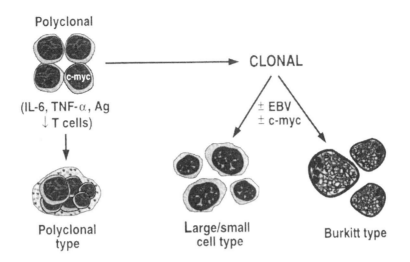

FIGURE 227

Emergence of B-cell lymphomas in HIV-infected individuals. Polyclonal proliferation of B cells can give rise to monoclonal lymphomas. The latter type may show evidence of *c-myc* translocation and/or EBV infection. The Burkitt type of lymphoma appears to develop from one malignant cell. EBV and *c-myc* may be detected in polyclonal lymphomas but at a reduced frequency. (From Levy, J.A., *HIV and the Pathogenesis of AIDS,* 2nd ed., ASM Press, Washington, 1994, 189. With permission by B.G. Herndier.)

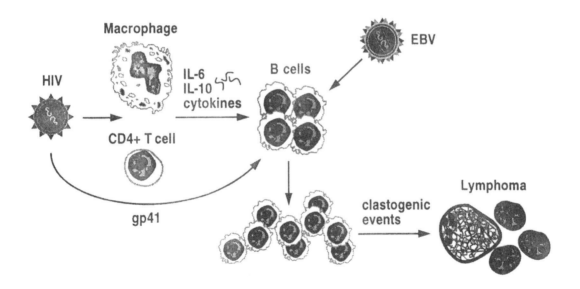

FIGURE 228

Potential mechanisms for transformation of B cells into lymphoma. HIV infection of macrophages or CD4+ T cells can result in the production of cytokines that induce the proliferation of B cells. The HIV gp41 envelope protein and EBV can also be involved. The polyclonal activation and chromosomal changes (e.g., c-*myc* translocation) cause the emergence of an autonomously growing malignant cell. (From Levy, J.A., *HIV and the Pathogenesis of AIDS,* 2nd ed., ASM Press, Washington, 1994, 189. With permission.)

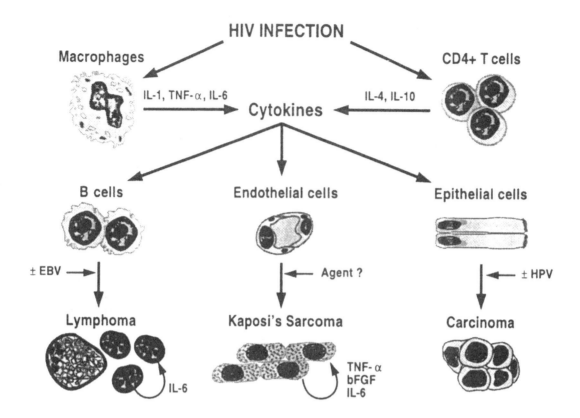

FIGURE 229

Induction of cancers by HIV. Virus infection of macrophages, CD4+ cells, or other cells could lead to the production of cytokines that enhance the production of certain target cells such as B cells, endothelial cells, and epithelial cells. Enhanced replication of these cells - either through apocrine cytokine production or through subsequent viral infection (EBV, HPV) - could lead to the eventual development of the malignancies noted. In some cases, such as B-cell lymphomas and Kaposi's sarcoma, ongoing cytokine production by tumor cells maintains the malignant state. (From Levy, J.A., *HIV and the Pathogenesis of AIDS*, 2nd ed., ASM Press, Washington, 1994, 198. With permission.)

6.XVIII. RHABDOVIRIDAE

Linear (-) sense ssRNA, nonsegmented
Helical, enveloped

This large, diversified family has five genera and an exceptional host range, occurring in vertebrates, invertebrates, and plants. Viruses of three genera infect vertebrates, notably members of the genera *Lyssavirus* (rabies) and *Vesiculovirus* (vesicular stomatitis). Particles are bullet-shaped (vertebrate and insect viruses) or bacilliform (many plant viruses), measure 100-430 x 50-100 nm, and consist of an envelope and a tubular nucleocapsid formed by a coiled filament of protein and RNA.

RABIES

Human rabies is a fatal acute encephalitis. Rabies virus is neurotropic and has a special affinity for salivary glands. Wild animals (fox, wolf, skunk, raccoon, bat) constitute the natural reservoir and transmit the virus to dogs and other domestic animals. In turn, dog and cat rabies is the main source of human rabies. The virus is generally transmitted by bite and inoculation of infected saliva. The incubation period, extremely variable, is usually 1 to 2 months. The virus remains first at the site of entry, replicates then in myocytes, and finally ascends via peripheral nerves to the CNS, from where it descends to the salivary glands and other locations. Clinical rabies, characterized by the pathognomonic symptom of hydrophobia, presents itself in a "furious" and, less frequently, paralytic form, and ends in coma and death.

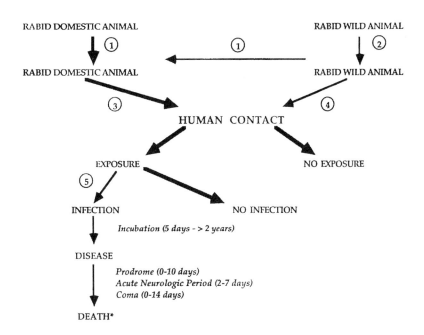

NATURAL HISTORY OF HUMAN RABIES

FIGURE 230

Natural history of human rabies and approaches to prevention and control. The chain of infection may be broken at any of the numbered points: (1) vaccination and control of domestic animals; (2) vaccination of wild animals; (3), avoidance of domestic animals; (4) avoidance of wild animals; and (5) pre- and postexposure prophylaxis (including local wound care, passive immunzation, and active immunization. (Reprinted with permission from Fishbein, D.B., in *The Natural History of Rabies*, 2nd ed., Baer, G.M., Ed., CRC Press, Boca Raton, 1991, 519. Copyright CRC Press, Boca Raton, Florida.)

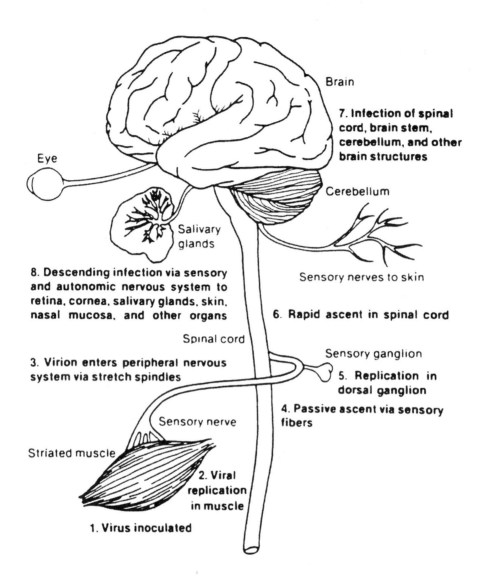

FIGURE 231

Pathogenesis of rabies virus infection. Numbered steps describe sequence of events of infection. (From Robinson, P.A., in *Textbook of Human Virology,* 2nd ed., Belshe, R.B., Ed., Mosby-Year Book, St. Louis, 1991, 517. With permission.)

6.XIX. TOGAVIRIDAE

Linear (+) sense ssRNA, nonsegmented
Cubic, enveloped

Togaviruses include the genera *Alphavirus* (type A arboviruses) and *Rubivirus* (human rubella virus). They infect mammals, birds, and insects. Particles are spherical, measure 60-70 nm in diameter, and consist of an envelope, an icosahedral capsid, and a single RNA molecule. Alphaviruses cause essentially the same types of disease as flaviviruses, namely fever with arthralgia and rash, hemorrhagic fever, and encephalitis. They are transmitted by mosquitoes and ticks. Sindbis virus is the type species. It causes acute encephalitis in mice and is a useful model for encephalitic alphavirus infections in humans. The outcome of Sindbis infection is age-dependent and is determined by the ability of viruses to induce apoptosis in infected neurons. Failure to induce apoptosis leads to long-term persistence of viral RNA within infected neurons.

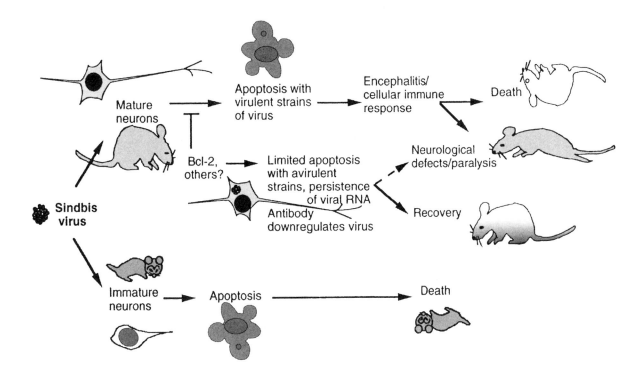

FIGURE 232

Age-dependent susceptibility to Sindbis virus in mice. The outcome is determined by whether infected neurons are resistant to virus-induced programmed cell death or activate their apoptotic pathway. The host immune response may also cause death of infected neurons. Determinants of neuronal apoptosis include the maturity of the neuron, the virulence of the infecting virus and the cellular immune response infection. In mature cultured neurons of older mice (>2 weeks), infection results in minimal apoptosis and establishment of persistent infections. Most animals recover from encephalitis with little sequelae. More virulent strains overcome the protective/antiapoptotic activity present in mature neurons, leading to apoptosis and death. In contrast, freshly explanted neurons in culture and newborn mice exhibit massive apoptosis and die. (From Griffin, D.E. and Hardwick, J.M., *Sem. Virol.,* 8, 481, 1998. With permission.)

RUBELLA

This human-specific disease, also called "German measles", is caused by the single member of the genus *Rubivirus*. Rubella is generally a childhood disease and is essentially benign in children and adults, causing lymphadenopathy, conjunctivitis, macular exanthem, and frequently arthritis. The virus is teratogenic. In nonimmunized women, the virus infects the fetus; when infection occurs in the first trimester of pregnancy, it often leads to fetal abnormalities (blindness, deafness, heart defects). Nonteratogenic effects range from growth and mental retardation to purpura, hepatosplenomegaly, pneumonia, and death in the first year of life.

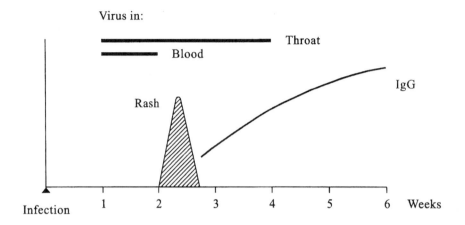

FIGURE 233

Course of rubella virus infection in children and adults. (By Ackermann.)

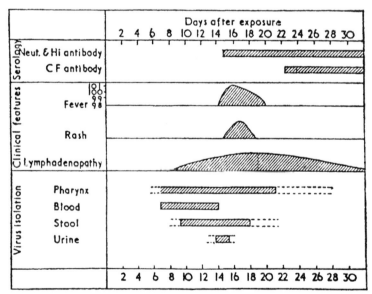

FIGURE 234

Relationship between clinical features, virus excretion and serological responses in postnatal infection with rubella virus. CF, complement fixing; Hi, hemagglutination-inhibiting. [Author's note: this historical diagram shows that rubella virus pathogenesis was well known at the end of the 1960s. An updated version has recently been published in *Principles and Practice of Clinical Virology,* 4th ed., Zuckerman, A.J., Banatvala, J.E., and Pattison, J.R., Eds., John Wiley & Sons, Chichester, 2000, 387.) (From Banatvala, J.E., *Brit . Med. J.,* i, 561-562, 1968. With permission of the BMJ Publishing Group, London.)

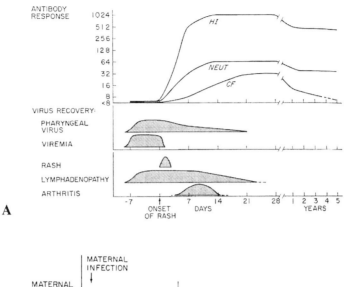

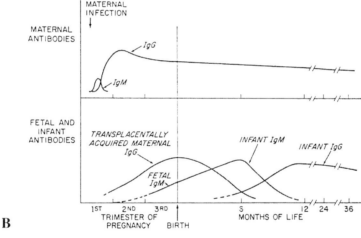

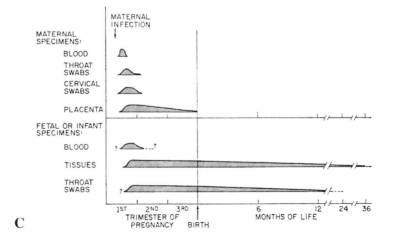

FIGURE 235

(A) Clinical and virologic features of natural rubella and sequence of immunologic responses (B) and virologic events (C) in maternal-fetal rubella. (Adapted from Meyer, H.M., Parkman, P.D., and Hopps, H.E., *Am. J. Clin. Pathol.*, 57, 803, 1972. With permission.)

Chapter 7

INSECT VIRUSES

Most insect viruses have been found in pests or beneficial insects such as bees, silkworms, and parasitic wasps. Many of them are viruses for large and easily observed insects such as lepidoptera. Since there are over 1 million insect species in nature and about 15 honey bee viruses alone, it is safe to conclude that insect viruses probably constitute the largest of all viral groups. Some insect viruses have acquired considerable importance in biological pest control and, in gene technology, as expression vectors for eukaryotic genes.

Similarly to vertebrate viruses, insect viruses are extremely varied in morphology and nature of nucleic acid. The presently known insect viruses belong to 14 families, 33 genera, and one unassigned group. The various insect virus families:

1. Are specific to insects *(Asco-, Baculo-, Polydna-, Tetraviridae, Oryctes* virus group).
2. Include genera specific to insects and others specific to vertebrates, plants, or yeasts *(Birna-, Irido-, Noda-, Poxviridae).*
3. Include genera replicating in both insects and vertebrates *(Bunya-, Flavi-, Reo-, Rhabdo-, Togaviridae).*
4. Include genera replicating in both insects and plants *(Bunya-, Reo-, Rhabdoviridae).*

The role of viruses "bridging" several host kingdoms in the origin and evolution of viruses can only be guessed. Clearly, insects play a central role in the viral world. At the present state of knowledge, it is impossible to say whether, for example, insect poxviruses originated in vertebrates or vice versa.

Insects are well protected against infections. The main line of defense is their thick and rigid integument, composed of the cuticle and the epidermis. The cuticle is a complex, multilayered exoskeleton with layers of cement, waxes, and chitin. The latter constitutes the bulk of the integument. The epidermis is generally a single layer of cells. No part of the insect is directly exposed to the outside because the cuticle even covers foregut, hindgut, and tracheal tubes. The midgut is guarded by a special membrane.

The inner line of defense against viruses consists of blood cells specialized in phagocytosis, the plasmocytes. Insects have no interferon, lymphocytes, complement, or immunoglobulins; there is no acquired immunity. Apoptosis of infected cells occurs in baculovirus infections and may constitute a way of defense. Their other defenses against foreign invaders, namely encapsulation, nodule formation, and production of lysozyme and phenoloxidases, seem to be directed against large cellular agents such as bacteria or fungi rather than against viruses.

Although insect viruses are able to cause devastating epizootics when host population density is high, the general pattern in nature seems to be one of equilibrium between insect and virus. Many viruses are host species-specific. In a general way, the symptoms of disease are nonspecific and give few indications on the agent of disease or its location within the host. Infected cells may contain enormous quantities of virus. Larvae are more susceptible than adults. Insects are infected by:

1. Piercing-sucking on vertebrates or plants.
2. Feeding on infected foliage or dead larvae.
3. Inoculation by parasitic wasps.
4. Vertical transmission.

Foliage may be contaminated by larval feces, regurgitated food, or decaying larvae. After infection by feeding, the incubation period depends on the age of the larva, the infecting dose, and environmental conditions (mainly temperature). Ingested viruses infect the insect via the columnar cells of the midgut or are probably destroyed in the gut lumen. Viruses are disseminated by rain, wind, birds, predators, parasites, and migratory larvae and adults. Remarkably, several types of widely distant insect viruses have evolved occlusion bodies for protection of viruses against the environment. These viral-encoded occlusion bodies are proteinic and intranuclear (baculoviruses) or intracytoplasmic (entomopoxviruses, *Cypovirus* genus of the *Reoviridae* family). They are extremely resistant against the environment; for example, the infectivity of baculovirus polyhedra persists over 240 weeks.

Infection by parasitic wasps is done by egg-laying into host larvae. Vertical transmission occurs in *Polydnaviridae* as they are transmitted as integrated provirus genomes from one wasp to another. Transovarial transmission seems to occur, but has not been conclusively proven in insects.

The pathogenesis of baculoviruses, the possibly related *Oryctes*-type viruses, and polydnaviruses has been illustrated by diagrams and is discussed here. Information on other insect viruses and "bridging groups" may be found in reference 192.

BACULOVIRIDAE infect insects and crustacea (shrimps). They are well known, partly because the polyhedral inclusion bodies produced by many of them are easily visible in the light microscope. Virions contain supercoiled circular dsDNA, consist of an envelope with one or more nucleocapsids of 220 to 400 x 50 nm, and are occluded in virus-specified protein crystals. Many viruses have narrow host ranges. The family has two genera:

The genus **Nucleopolyhedrovirus** is characterized by large polyhedra of 1-15 μm in size containing 20 to 200 viruses of two phenotypes, one with a single and the other with multiple (up to 32) nucleocapsids per envelope. Their complicated replication cycle (Fig. 239) has two phases. Ingested polyhedra are dissolved by alkaline gut juices (pH >10) in the midgut. Free enveloped viruses enter midgut epithelial cells, replicate in their nuclei, and are released by budding into the hemocoel. In the second phase, these "budded viruses" infect blood and tracheal cells, and the fat body (other cells may be infected later). Newly formed viruses replicate in the nuclei, are enveloped there, and are embedded in a protein matrix, the future polyhedra. Host protein synthesis is shut down and nuclei are fragmented. Death occurs after 4 days to 3 weeks. Dying larvae are so full of polyhedra that they break easily and liberate liquefied body contents.

The genus **Granulovirus** is characterized by the production of small oval "granules" of 0.3 to 0.5 μm in length which contain one virus with a single nucleocapsid and found mainly in fat body and epidermis. Virus replication is essentially as above, except that the nuclear membrane disintegrates early in infection and that nucleocapsid assembly takes place in a mixture of nuclear and cytoplasmic elements.

The **Oryctes** **virus group** is found in insects, including the scarab beetle, *Oryctes rhinoceros,* as well as in arachnids and crustacea (crabs). Viruses resemble baculoviruses in morphology and nature of DNA, but are generally shorter and nonoccluded. They were for a time classified as members of the *Baculoviridae*, but are now unassigned to any family. In *Oryctes* beetles, the ingested virus spreads from the hindgut into the hemocoel and the fat body. Adults and larvae are sensitive. The virus is a very useful agent for control of *Oryctes* beetles, which are a widespread and devastating pest of coconut palms.

POLYDNAVIRIDAE are unusual in many ways. They are the only dsDNA viruses with segmented genomes, they are specific and beneficial to parasitic wasps, and they are transmitted in these vertically as proviruses. Virions are enveloped and have cylindrical nucleocapsids. *Bracovirus* nucleocapsids are of variable length (30-300 x 40 nm) and may have taillike appendages. Envelopes include several nucleocapsids. By contrast, virions of the genus *Ichnovirus* have a single fusiform nucleocapsid of 330 x 85 nm surrounded by two envelopes. Viruses are produced in the oviduct of female wasps and are introduced with the egg into caterpillars (Fig. 243). Infected larvae normally defend themselves by encapsulation of parasite eggs, but polydnaviruses suppress this encapsulation reaction.

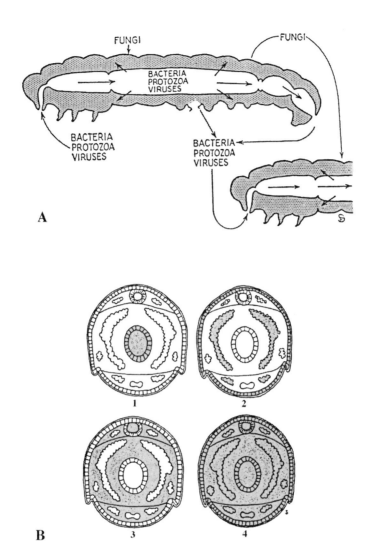

FIGURE 236

(A) Routes by which infectious microorganisms can gain entrance into an insect host and be disseminated again. (B) Types of infection occurring in insects as visualized in cross-sections. (1) The intestinal, or dysentery, type of infection in which the invading organism is limited to the alimentary tract and its appendages. (2) Tissue infection, here indicated by the stippling of the adipose tissue and the hypodermis. (3) Septicemic type of infection in which the invading organism multiplies in, and is distributed throughout, the body cavity by the hemolymph. (3) General systemic infection in which the invading organism penetrates all parts of the insect's body. (Adapted from Steinhaus, E.A., *Principles of Insect Pathology,* McGraw-Hill, New York, 1949, 168 and 172.)

7.I. BACULOVIRIDAE

Supercoiled dsDNA
Enveloped, helical

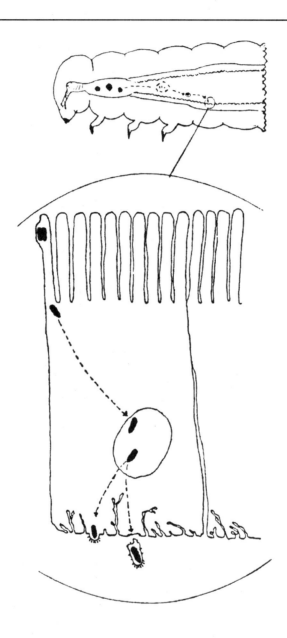

FIGURE 237

The natural infection cycle of baculoviruses begins when viral inclusion bodies are ingested by susceptible larvae. The occluded virus is released in the midgut by dissolution of the inclusion bodies by alkaline gut juices. The released virus infects the columnar epithelial cells of the midgut. Viral replication takes place in the nucleus of these cells, and progeny virus (of a different phenotype than the originally infecting virus) buds from the basement membrane of the hemocoel. This budded virus causes a systemic infection with production of both the occluded and budded viral phenotypes. (Fig. 1 from Volkman, L.E., in *Curr. Topics Microbiol. Immunol.,* 131, 103, 1986. © Springer-Verlag. With permission.)

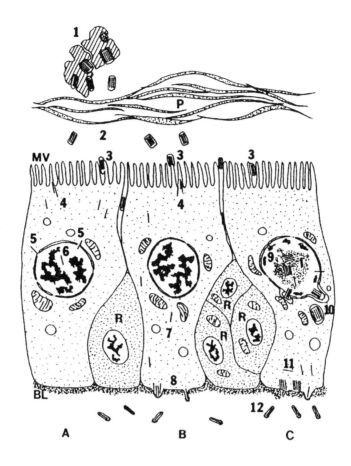

FIGURE 238

Infectious pathways of NPV in insect larvae. Occlusion bodies (OB) are ingested and dissolved in the midgut of the larvae, releasing the embedded virions (1). Released virions pass the peritrophic membrane (P) and associate with columnar epithelial cells of the midgut wall (2). Virions contact microvilli (MV) or lateral surfaces of these cells and gain entry by membrane fusion (3). Once inside the cell, two routes may be followed: initiation of cell replication (cell A) or translocation to the basal membrane (cell B). R, epithelial regenerative cells. Cell A: Nucleocapsids are transported through the cytoplasm to the nucleus (4), probably through association with microtubules. They enter the nucleus probably through nuclear pores (5). Uncoating of the viral genome and initiation of viral replication occur in the nucleus (6). Cell B: In some cases nucleocapsids bypass the nucleus and migrate directly to the basal membrane of columnar epithelial cells (7). They can bud through basal membrane and basal lamina (BL) into the hemolymph, acquiring a cell membrane-derived envelope in the process. Cell C: Virus replication in the nucleus of columnar epithelial cells produces progeny nucleocapsids but no OB's. A diffuse virogenic stroma is formed in which nucleocapsid assembly takes place. (9) Nucleocapsids exit the nucleus by leaving through the nuclear envelope, forming transport vacuoles (10). The latter are lost during transport of nucleocapsids to the basal membrane (11) where budding of virus progeny takes place. Extracellular virus in the hemolymph (12) transmits the disease throughout the larval body. (From Fraser, M.J., *Ann. Entomol. Soc. Am.*, 79, 773, 1986. With permission.)

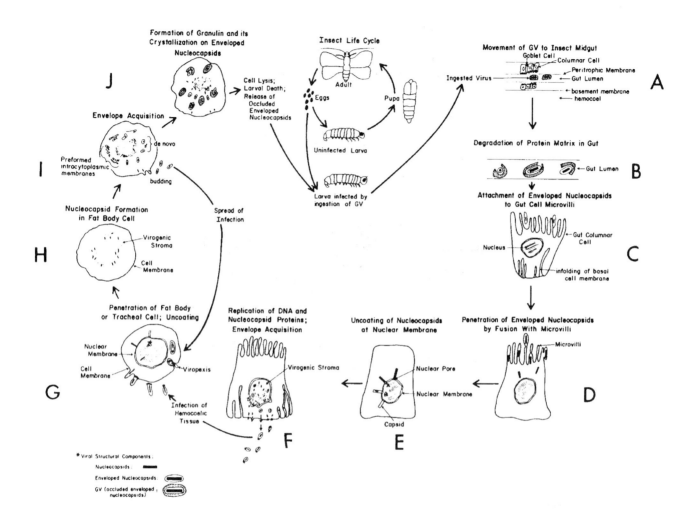

FIGURE 239

The granulosis virus (GV) infection process. (A) Larvae ingest granulosis inclusion bodies. (B) Enveloped nucleocapsids are released by digestion of inclusion body protein and (C, D) attach to and penetrate cells of microvilli. (E) Nucleocapsids are uncoated at the nuclear membrane. (F) Viruses replicate within the nucleus, are assembled, leave the cell by budding, and infect cells of the hemocoel (G) and fat body or tracheal cells (H). Novel viruses acquire envelopes by *de novo* synthesis or from preformed cellular membranes (I) and may return to the hemocoel or (J) become occluded by crystallization of granulin on viral envelopes. The larva now dies and virus-containing granules are liberated into the environment. (From Tweeten, K.A., Bulla, L.A., and Consigli, R.A., *Microbiol. Rev.*, 45, 379, 1981. With permission.)

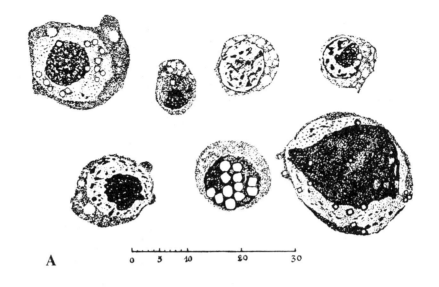

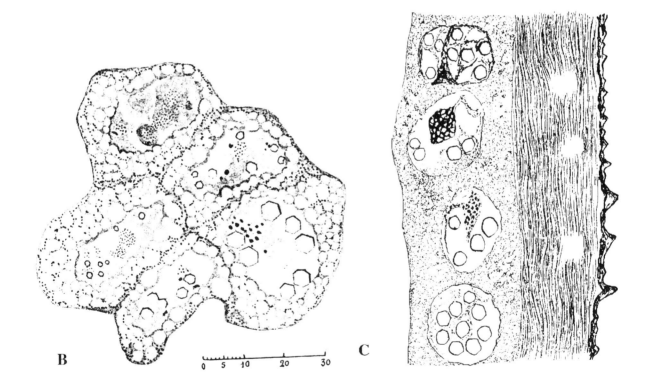

FIGURE 240

Observations in the *grasserie*-infected silkworm. (A) Blood cells; Giemsa. (B) Fat cells; formol-salt fixation, Kull staining. (C) Sectioned hypodermis; fixation after Duboscq-Brasil, hematoxylin. Scale is in µm. (Adapted from Paillot, A., *L'Infection chez les Insectes. Immunité et Symbiose,* Imp. G. Pâtissier, Trévoux, 1933, 98 and 99. With permission by Éditions de Trévoux - Vincent Pâtissier, Route de Vienne, 69007 Lyon, editions.de.trevoux@wanadoo.fr)

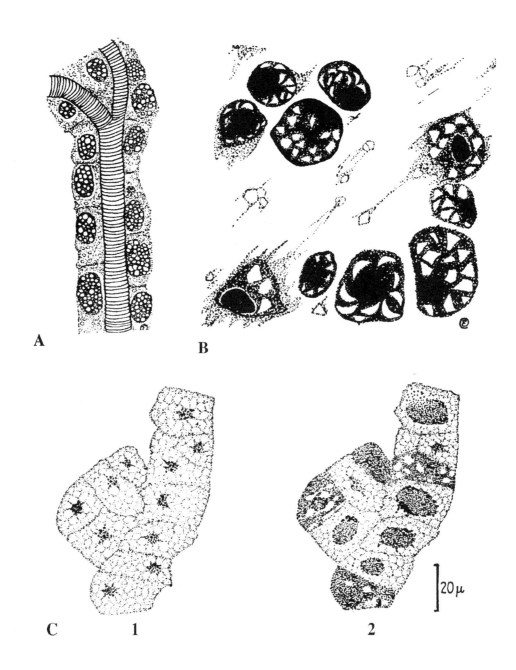

FIGURE 241

Polyhedra and granules in various insects. (A) Part of a tracheal tube of a diseased gypsy-moth *(Porthetria dispar)* caterpillar showing the presence of polyhedra in the cell nuclei of the tracheal matrix. (B) Polyhedral characteristic of the polyhydrosis of the larva of *Tipula paludosa* (Meig.), as they appear in the hypertrophied nuclei of fat cells. The cytoplasm has largely disappeared. The dark areas in the nuclei represent chromatin masses. (C) Granulosis type 2 of *Euxoa segetum* Schiff. (1) Portion of adipose tissue from a normal *Euxoa* larva. (2) The same portion showing pathological changes. Note hypertrophied and disintegrating nuclei and the characteristic small granular inclusions. (After photomicrographs by Glaser, W., *J. Agr. Res.*, 4, 101, 1915; Paillot, A., *Ann. Epiphyt. Phytogenet.*, 2, 341, 1936; Rennie, *Proc. Roy. Phys. Soc.*, A20, 265, 1923); from Steinhaus, E.A., *Principles of Insect Pathology*, McGraw-Hill, New York, 1949, 462, 494, and 505.) [With permission of the Royal Society of Edinburgh (B) and INRA (C).]

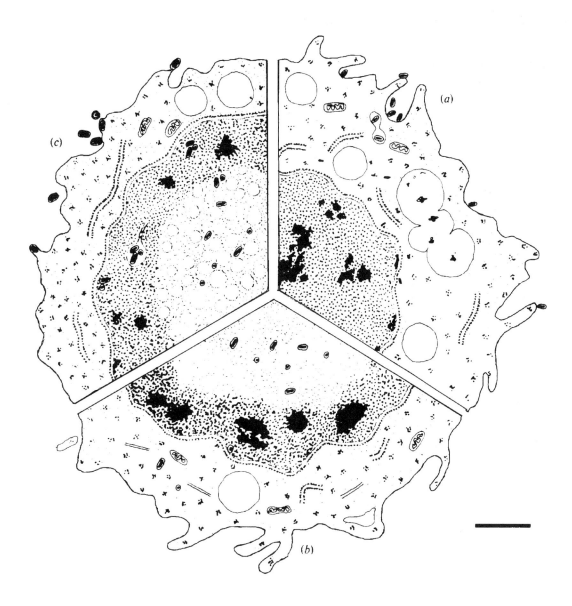

FIGURE 242

Morphological changes in HA cells infected with *Oryctes* baculovirus. Cells take up virus particles by pinocytosis at the plasma membrane. Seven h after infection, enveloped virus particles appear in a clear area in the nucleus. Virus is released by budding through the plasma membrane and a second unit membrane is acquired in the process. The infectivity of viruses with a single or double envelope is essentially identical. (a) 1 to 4 h post-infection: virus adsorption to the plasma membrane and uptake in cytoplasmic vesicles. (b) 7 to 12 h post-infection: viral replication in the clear area of the enlarged nucleus. (c) 16 and over h post-infection: virus release by budding and characteristic accumulation of envelope material in the nucleus. The bar represents 1 μm. (From Crawford, A.M. and Sheehan, C., *J. Gen. Virol.*, 66, 529, 1985. With permission.)

7.II. POLYDNAVIRIDAE

Supercoiled segmented dsDNA
Enveloped, helical

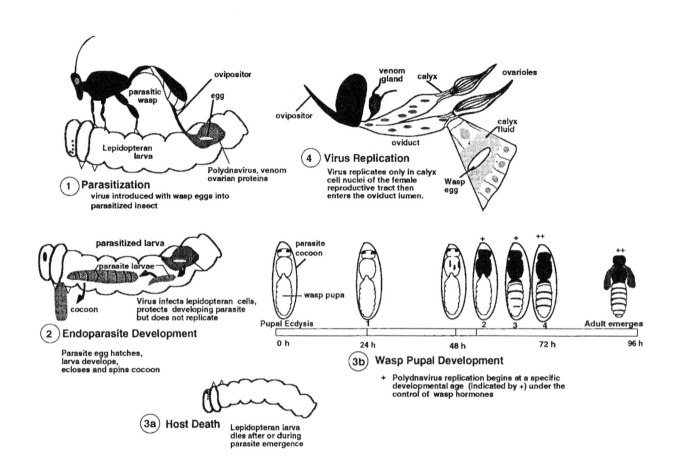

FIGURE 243

Polydnavirus life cycle. Virus particles are injected into a lepidopteran larva along with a wasp egg during oviposition, the first step in parasitization (1) of the larva by the wasp. The wasp carries and vertically transmits the polydnavirus in proviral form. The egg develops within the parasitized larva (2) with the prospective aid of the polydnavirus. The lepidopteran larva dies (3a) during or after parasite emergence. As the wasp develops (3b), the amplification of polydnavirus DNA in calyx cells of the female oviduct is first observed at a specific stage in wasp pupal development (ca. 48 hr after pupal ecdysis). The amplified polydnavirus DNA is packaged into viral particles which are released from epithelial cells in the calyx of adult female ovaries (4). The wasp egg is bathed in calyx fluid which contains polydnavirus particles. The fluid and wasp egg are injected, along with venom gland secretions, upon oviposition (1). (Drawing by B.A. Webb, University of Kentucky; from Miller, L.K., in *Fields Virology*, 3rd ed., Vol. 1, Fields, B.N., Knipe, D.M., and Howley, P.M., Eds.-in-chief, Lippincott-Raven, Philadelphia, 1996, 533. With permission by B.A. Webb.)

Chapter 8

PLANT VIRUSES

Plant viruses belong to 13 families and about 70 genera, many of which have not been assigned to a family. They comprise viruses with single- or double-stranded RNA or DNA and enveloped viruses such as bunya- and rhabdoviruses, but most appear remarkably uniform. About 75% of taxa contain single-stranded (+) RNA and the vast majority are nonenveloped icosahedra, rods, or filaments. Moreover, icosahedral particles are generally 28-30 nm in diameter and have T=3 capsids. It is noteworthy that the plant viruses with dsDNA use reverse transcriptase for replication. Another particular feature is the occurrence of multipartite systems with 2, 3, or 4 different RNA molecules each packaged into a different capsid.

Virus diseases can cause catastrophic economic losses in important cultivated plants, for example cassava, citrus trees, cocoa plants, corn, or sugar beets. The members of the large *Potyviridae* family are among the most destructive. In a general way, the disease depends on the age and physiology of the host, viral strains, climate, and the environment. Host range is extremely variable. Alfalfa mosaic virus can infect over 430 plant species in 50 plant families, but other viruses are limited to a single host species.

Plants are protected against dessication and injury by exceptionally thick cell walls overlaid with a film of waxes, a mixture of polymeric fatty acids called cutin, and pectins. Cell walls are rigid and consist of cellulose fibrils embedded in a matrix of hemicellulose and pectins. This barrier must be breached to achieve infection, which explains the importance of vectors in plant virology. There is no simple contamination by air or water. Plant viruses are transmitted from plant to plant:

1. Mechanically in nature through broken leaf hairs rubbing together or by damage inflicted by animals (dogs, rabbits).
2. Mechanically by man in harvesting, grafting, pruning, or, in plant virology, rubbing plant surfaces with an instrument or abrasive.
3. Infected bulbs, tubers, pollen, or seeds.
4. Biological vectors, generally piercing-sucking or biting insects. Vectors include insects, mites, nematodes, and fungi.

Aphids are probably the most important insect vectors, followed by plant and leaf hoppers, whiteflies, mealybugs, thrips, and beetles. The preferred food is the nutrient-rich phloem sap. Roots may be attacked by soil-inhabiting nematodes which feed like piercing insects, and by lower fungi such as *Olpidium brassicae,* which carries viruses attached to motile zoospores. The relationship between a virus and its vector is often very specific. Unlike animal or bacterial viruses, viruses are not internalized after adsorption to specific receptors. According to their association with a vector, plant viruses are:

1. Nonpersistent if retained by the vector for a few hours only.
2. Semipersistent if retained for a few days.
3. Persistent if retained by the vector for the rest of its life.

Nonpersistent and semipersistent viruses do not enter the hemocoel of the vector and do not cross vector cell membranes. They are released by salivation and termed noncirculative. By contrast, persistent viruses are circulative pass, through the hindgut into the hemocoel, cross vector cell membranes, and return to the salivary glands. The capsids of some circulative viruses, e.g., luteoviruses and enamoviruses, contain small amounts of a "readthrough protein" that seems to be required for access to salivary glands. Circulative viruses may be propagative, replicating in vectors and plants, and nonpropagative, replicating in plants only. Propagative viruses include the *Tospovirus* genus of the *Bunyaviridae* and all plant virus genera of the *Reoviridae* and *Rhabdoviridae* families.

Inside the plant, viruses spread from cell to cell or through the vascular system. They may be inoculated directly into the nutrient-carrying phloem or the water-carrying xylem. Cell-to-cell or short-distance movement is slow. Viruses move through plasmodesmata, a system of intercellular pores and channels called the "symplast" and lined by plasma membrane. Viruses move as complete particles or, possibly, nucleoprotein complexes, free RNA, or replicative intermediates. Viral spread is mediated by virus-encoded movement proteins. One mechanism is exemplified by tobacco mosaic virus (TMV) and the other by cowpea mosaic virus (CPMV). In TMV, a 30 KDa nonstructural protein binds to viral RNA and increases plasmodesmata diameters. In CPMV, entire virus particles move through virus-carrying tubules with the help of capsid and nonstructural virus proteins.

Long-distance movement is rapid and occurs essentially in the phloem in cells connected by sieve plates with large pores. Again, viral spread is mediated by virus-encoded proteins. The viral genes required for short- and long-distance movement seem to be different. In the case of TMV, the coat protein is involved in long-distance transport of the virus.

The pattern of virus movement, first investigated in a classical study by Samuel (Fig. 249) depends on the flow of the products of photosynthesis, for example from mature leaves to immature leaves, and on leaf arrangement. It may comprise return flows. The walls of phloem cells may constitute barriers for entry or exit of viruses to or from parenchyma cells.

Typically, plant viruses cause generalized, persisten infections. The infected cells may contain enormous quantities of virus, but the plants are usually not killed by the virus, do not eliminate it, and remain infectious.External symptoms include mosaics, mottling, yellowing and other color changes, vein clearing, necrotic spots and ringspots, wilting, swelling and tumors such as galls, dwarfism, and slowed reproduction. Mosaics are patterns of dark-green and light-green or yellow areas. Vein clearing is vein yellowing caused by virus moving intoleaves. Necroses are hypersensitivity reactions with synthesis of nonspecific "pathogenesis-related proteins" (PR). The infected cells die and the infection becomes self-limiting. Ringspots are concentric circles of dead cells alternating with normal cells.

Internal symptoms include histological changes such as local hyperplasia, hypertrophy, hypoplasia, and necrosis. Vein clearing is due to hypertrophy of cells adjacent to veins. Cytological changes affect the arrangement and appearance of cell components (nucleus, chloroplast, mitochondria); for example, tymoviruses induce invaginations of the chloroplast membrane. Cells frequently contain intracytoplasmic (less often intranuclear) inclusion bodies which may be amorphous, fibrous, banded, crystalline, sheet- or needle-like. Potyviruses induce unique whorl-like structures.

Plant virology has yet to solve many problems. Although some movement proteins have been identified, it is unclear if this is a general feature, in which form viruses move from cell to cell, and how they move in and out of the phloem. Similarly, information on the molecular basis of host range and on molecular pathogenesis is still scarce. For more information on plant virology, the reader is referred to the textbook by R.E.F. Matthews[199] and several recent reviews. [200-203]

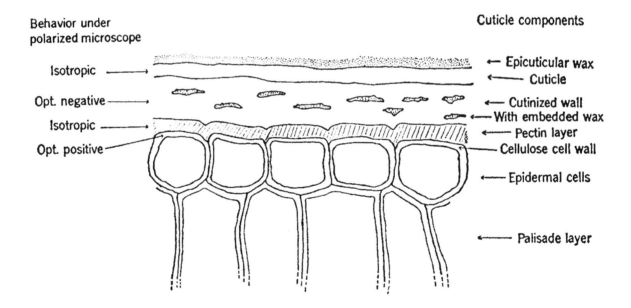

FIGURE 244
The plant epicuticle in cross-section. The lines dividing the layers above the epidermal cells indicate regions of major change in the construction of components rather than sharp boundaries. Individual plant species may greatly depart form this general arrangement. (Reprinted with permission from Eglinton, G. and Hamilton, R.J., *Science*, 156, 1322, 1967; source B.E. Juniper, Botany School, Oxford University. Copyright 1967, American Association for the Advancement of Science.)

8.I. INFECTION AND VIRUS SPREAD

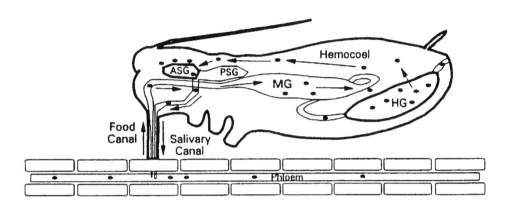

FIGURE 245

Internal structures of aphids that are involved into the circulative transmission of luteovirus. The circulative pathway, indicated by arrows, requires that the virus be actively transported from the hemolymph into the salivary system. It appears the luteovirus read-through domain is required for virus transport through the membranes of aphid salivary glands, but is not required for systemic infection of plants and influences virus accumulation in plants. A viral nonstructural 17-kDa protein is required for systemic infection. ASG, accessory salivary gland, HG, hindgut; MG, midgut; PSG, principal salivary gland. (From Chay, C.A., Gunasinge, U.B., Dinesh-Kumar, S.P., Miller, W.A., and Gray, S.M., *Virology,* 219, 57, 1996. With permission.)

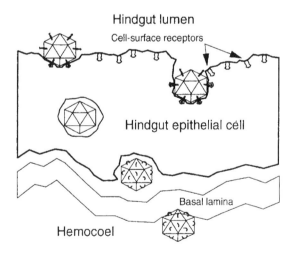

FIGURE 246

Proposed receptor-mediated endocytotic pathway of circulative nonpropagative viruses across hindgut epithelial cells. Accessible or protruding capsid protein domains bind to receptors on the apical plasmalemma of hindgut epithelial cells. The process of virus uptake or the environment within cytoplasmic transport vesicles might strip off the receptor-binding domain. The vesicle membrane fuses with the basal plasmalemma of the cell and virus is released into the hemocoel. (Reprinted from *Trends Microbiol.,* 4, 259-264, 1996. Gray, S.M., Plant virus proteins involved in natural vector transmission. © 1996, with permission of Elsevier Science.)

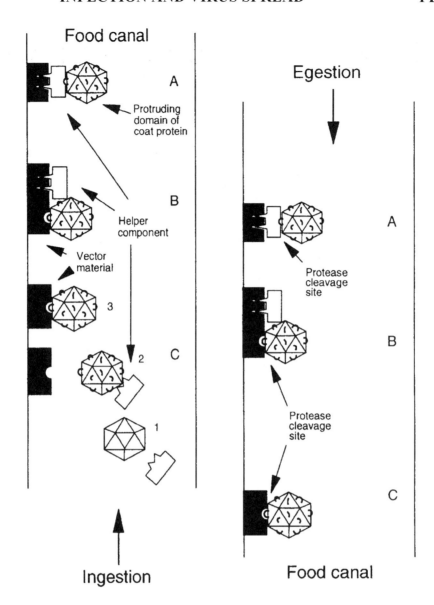

FIGURE 247

Three models of interaction between a noncirculative virus, the virally encoded helper component and vector material lining the cuticle in the food canal. Most experimental evidence supports either model **A** or model **B**. In model **A**, the helper component directly facilitates virus binding by first attaching to sites in the food canal; the virus then binds to the helper component. In **B**, the helper component indirectly facilitates binding of virus directly to the cuticle. Helper component first binds to a specific site causing a conformational change that allows virus to bind. In **C**, helper component interacts directly with the virus causing a conformational change in the virus. This exposes sites on the virus that can interact directly with binding factors on the cuticular lining of the food canal. The release of the virus, regardless of the binding process, occurs during the intermittent phases of feeding when the vector expels digestive secretions into the plant. The model proposes that proteases in the digestive secretions expose sites on the virus capsid that are involved in binding the virus to a helper component or to the vector. (Reprinted from *Trends Microbiol.*, 4, 259-264, 1996. Gray, S.M., Plant virus proteins involved in natural vector transmission. © 1996, with permission of Elsevier Science.)

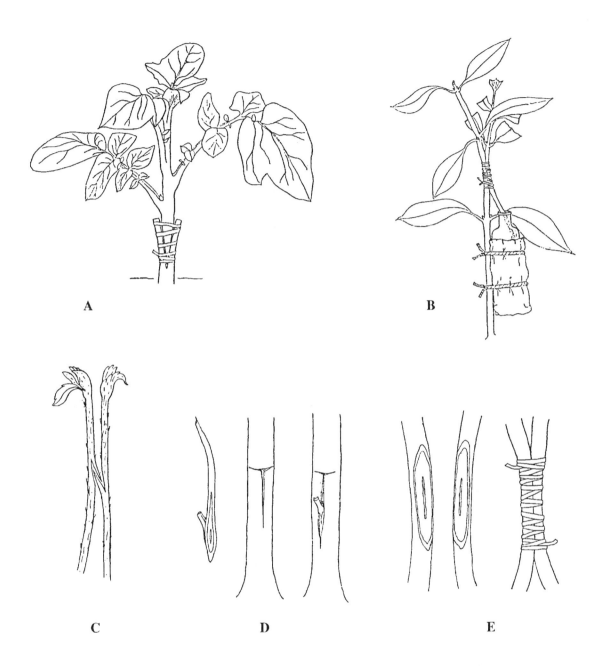

FIGURE 248

Types of graft used for virus transmission. (A) Wedge graft, used with herbaceous plants such as tomato; the cuts are made with a razor or thin sharp knife to avoid bruising the delicate stems. (B) Bottle graft; the base of the scion is kept in water until a union is formed and is removed later. (C) Tongue-cut approach, used with strawberry runners; the graft is later fitted with crepe rubber (not shown). (D) Shield bud graft, used with woody plants.   The bud (eye) is inserted under the bark of the stock ready for tying. (E) Spliced-approach graft. The components are sliced to expose the cambium in equal patterns. The shoots of rooted plants are pared to expose their cambium and the cut surfaces bound together. (After an arrangement by Gibbs, A. and Harrison, B., *Plant Virology, The Principles,* John Wiley & Sons, New York, 1976, 35. Adapted from Garner, R.J., *The Grafter's Handbook,* 3th ed., Faber and Faber, London, 1967, 103, 106, 110, 115, and 151. With permission by Mrs. I.L. Garner.)

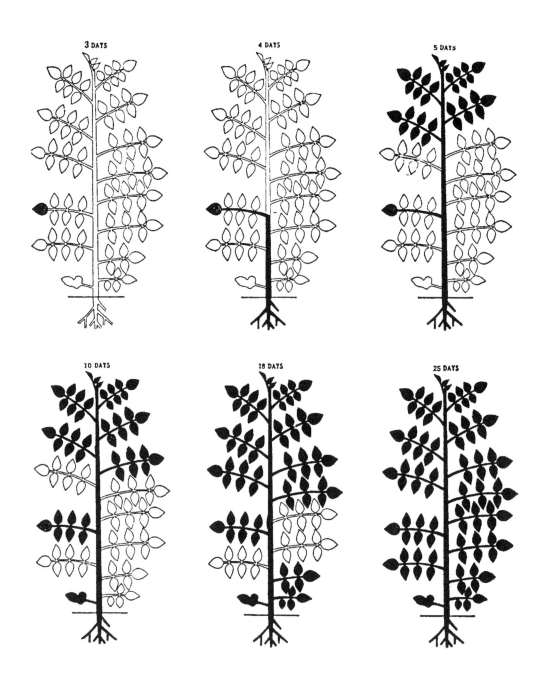

FIGURE 249
Progression of tobacco mosaic virus (in black) through a medium-sized, young tomato plant. Based on tests of Dwarf Champion plants about 15 in. high, growing in 6 in. pots in an unheated greenhouse. (From Samuel, G., *Ann. Appl. Biol.,* 21, 90, 1934. With permission.)

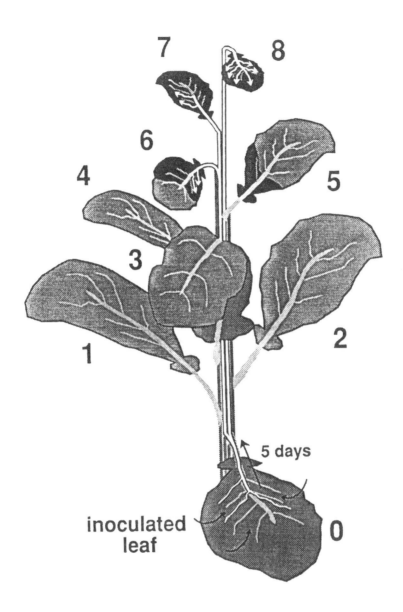

FIGURE 250
Turnip plant inoculated with cauliflower mosaic virus (CaMV) showing parameters influencing bilateral and basipetal accumulation of virus in leaves of systemically infected plants. The virus moves over long distances in plants within the phloem vasculature, i.e., from an older source leaf that exports photoassimilates to younger leaves. Leaves and leaf sectors that export photoassimilates are shaded. The inoculated leaf is indicated as leaf number 0. Black areas indicate immature (sink) parts of a leaf. (From Leisner, S.M., Turgeon, R., and Howell, S.H., *Mol. Plant Microbe Interact.*, 5, 41, 1992. With permission.)

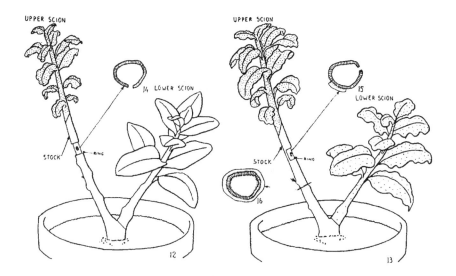

FIGURE 251

Translocation of curly-top virus in tobacco. Fig. 12. and 13. Two tobacco plants, each produced by grafting scions of *Nicotiana tabacum*, to a stock of *N. glauca*. The stem of the stock was "ringed" by removing bark in a ring-like area. Pith and internal phloem were also removed through a small hole. In Fig. 13 the ring was incomplete since some bark with phloem remained. In each plant the upper scion was inoculated with the virus and developed symptoms. In Fig. 12 the virus was unable to pass into the lower scion which remained healthy. In Fig. 13 the narrow phloem bridge allowed the virus to pass into the lower scion which developed symptoms. Fig. 14 and 15, cross-sections of the ringed parts of the stems in Fig. 12 and 13, respectively; xylem is indicated by hatching. Fig. 16, section of complete stem. (From Esau, K., *Am. J. Bot.,* 43, 739, 1956. With permission.)

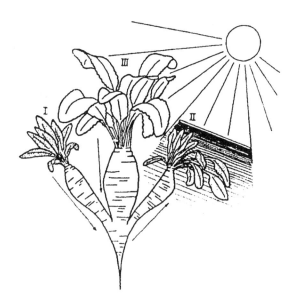

FIGURE 252

Translocation of curly-top virus in a sugar-beet plant with three crowns. Crown I was inoculated with the virus and developed symptoms. Crown II was shaded but not inoculated and developed symptoms. In the absence of light it did not form food and received its supply of food from the root. Virus appeared to have entered crown II from the root. Crown III was not treated in any way and remained free of symptoms. Arrows indicate prevailing direction of food movement. (Diagram by M. Shenkovsky, from Esau, K., *Am. J. Bot.,* 43, 739, 1956. With permission.)

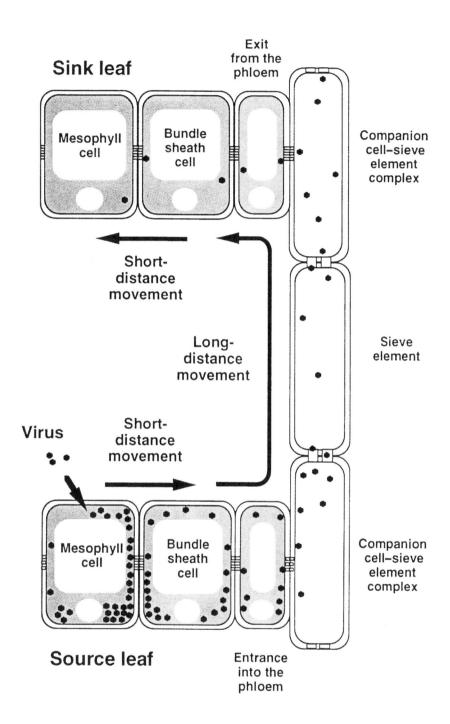

FIGURE 253

Flow-pattern of virus movement to sink leaves. Viruses (black hexagons) move short distances through nonvascular cells and long distances through phloem vascular cells. They move from cell to cell through plasmodesmata in vascular tissues and encounter a number of different cell types when entering and exiting the conductive (sieve element) cells of phloem vasculature. (Reprinted from *Trends Microbiol.*, 1, 314-317, 1993. Leisner, S.M. and Howell, S.H., Long-distance-movement of viruses in plants. © 1993, with permission of Elsevier Science.)

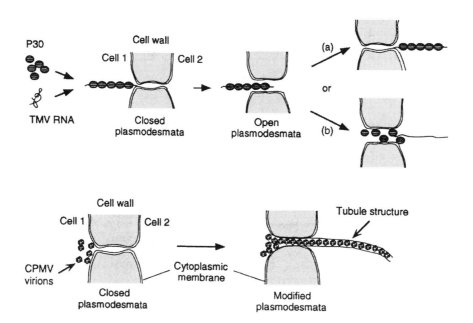

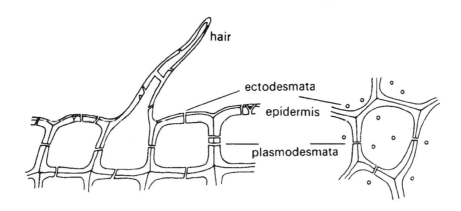

FIGURE 254

Two mechanisms by which virus-encoded movement proteins mediate cell-to-cell spread of viruses. The TMV-like mechanisms *(above)* involve transient modification of plasmodesmal width to allow transport of either (a) the ribonucleoprotein complex or (b) free RNA to the adjacent cell. The CPMV-like mechanism *(below)* involves formation of a tubule through which virions are transported to the adjacent cell; this tubule appears to modify plasmodesmal structure permanently. (Reprinted from *Trends Microbiol.,* 1, 105-108, 1993. McLean, B.G., Waigmann, E., Citowsky, V., and Zambryski, P., Cell-to-cell movement of plant viruses. © 1993, with permission from Elsevier Science.)

FIGURE 255

Ectodesmata - possible routes for virus entry. (From Stevens, W.A., *Virology of Flowering Plants,* Blackie & Sons, Glasgow, 1983, 95. With permission of Kluwer Academic Publishers.)

8.II. HISTO- AND CYTOPATHOLOGY

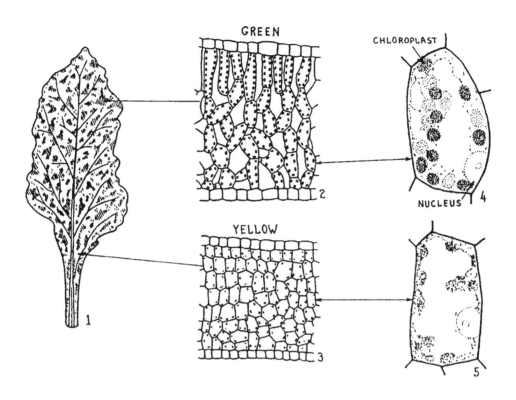

FIGURE 256

Effect of mosaic disease upon the sugar beet leaf. (1) Mosaic pattern on leaf. Green areas are shaded, yellow areas are left blank. (2) Mesophyll from a green area. It shows a loose arrangement of cells and numerous chloroplasts. (3) Mesophyll from a yellow area. It shows compact arrangement of cells like a young leaf. This underdevelopment is one of the expressions of hyperplasia. The chloroplasts are few. The deficiency in chloroplasts makes the tissue appear yellow. (4) Cell from green mesophyll with numerous chloroplasts. (5) Cell from yellow mesophyll. The chloroplasts have become partly or completely disorganized. (From Esau, K., *Am. J. Bot.,* 43, 739, 1956. With permission.)

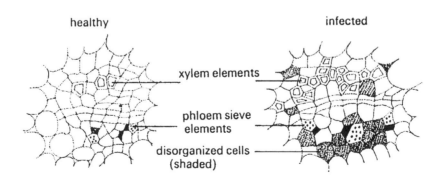

FIGURE 257

Transverse sections of tobacco leaves showing xylem proliferation and cell enlargement (hypertrophy) resulting from virus infection. (From Stevens, W.A., *Virology of Flowering Plants,* Blackie & Sons, Glasgow, 1983, 28. With permission of Kluwer Academic Publishers.)

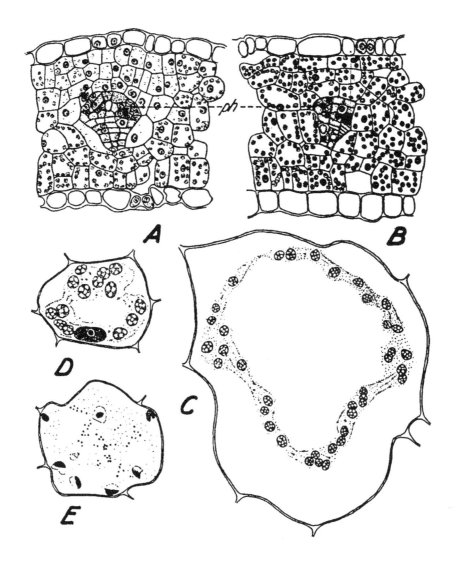

FIGURE 258

Clearing of veins. (A) Transverse section of young diseased leaf, showing hyperplasia of cells and degeneration of chloroplasts in mesophyll adjacent to phloem *(ph)*. (B) Transverse section of a young healthy leaf. (C) Parenchyma cell from a large vein of a young healthy leaf. (D). Healthy mesophyll cell from region adjacent to phloem of small bundle. (E) Diseased mesophyll cell from region adjacent to phloem of small bundle, showing degenerated chloroplasts. Original magnification: A and B, x 450; C-E, x 1060. (From Esau, K., *Phytopathology,* 23, 679, 1933. With permission.)

FIGURE 259
Vascular anatomy of a galled vascular bundle in a sugarcane leaf infected with Fiji disease virus. Only the distribution of xylem and phloem tissues is shown and a section of the bundle through the galled area has been removed to expose the tissues in transverse section. The galls appear to result from virus-induced cell proliferation. The proliferating cells develop into abnormal phloem (gall-phloem) and xylem (gall-xylem). Virus particles and viroplasms are confined to these tissues. (From Hatta, T., and Francki, R.I.B., *Physiol. Plant Pathol.,* 9, 321, 1976. With permission.)

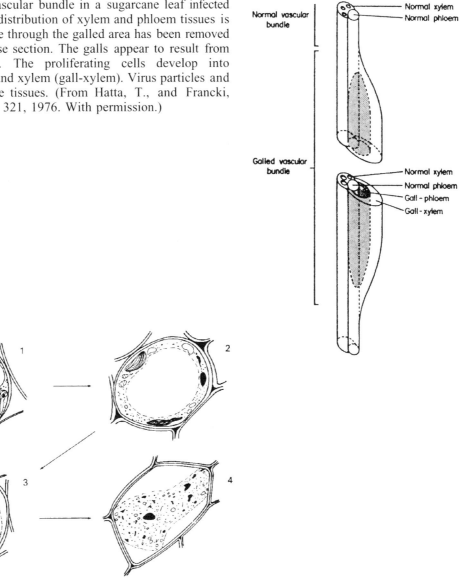

FIGURE 260
Changes in French bean leaf cell ultrastructure during local lesion production by TMV. (1) Healthy leaf mesophyllic cell. (2) Initial stages of infection. (3) Cell destruction in region near local lesion. (4) Complete loss of organelles in necrotic area of lesion. (After photomicrographs by Spencer, C. and Kimmins, W.C., *Can. J. Microbiol.,* 47, 2049, 1969; from Stevens, W.A., *Virology of Flowering Plants,* Blackie & Sons, Glasgow, 1983, 29. With permission of Kluwer Academic Publishers.)

Disease	Description	Diagrammatic Illustration
A chloroplast from a noninfected plant	The size, shape and distribution of lamellae are normal.	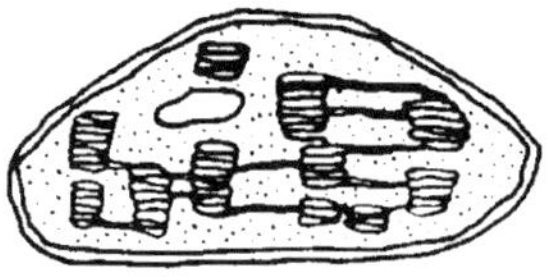
Barley Stripe mosaic virus in young barley plants	Chloroplasts lack a developed grana system. They primarily have intergrana lamellae.	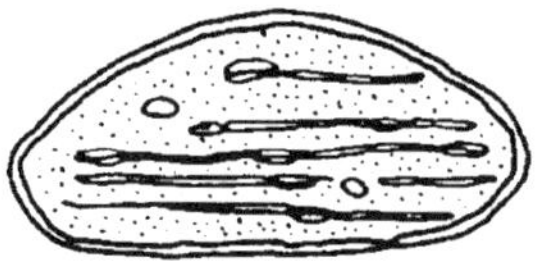
Arabis mosaic virus in *Arabis sp.* Southern bean mosaic virus in bean	Chloroplasts hump out at one or more places. There is an increase in stroma and no lamellae in the hump.	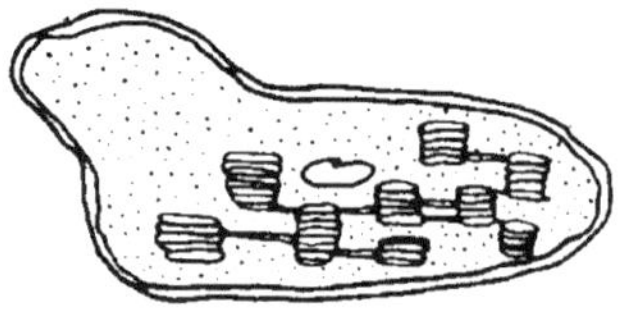
Some strains of tobacco mosaic virus in tobacco	An extrusion of the chloroplast which invaginates some of the ground cytoplasm of the cell. Mitochondria may even be found in the enclosure.	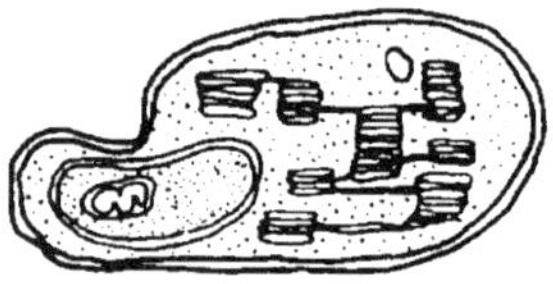
Turnip yellows mosaic virus in Chinese cabbage	Clumping of chloroplasts with vesicles in the center of the array. Chloroplasts have a limited system of lamellae.	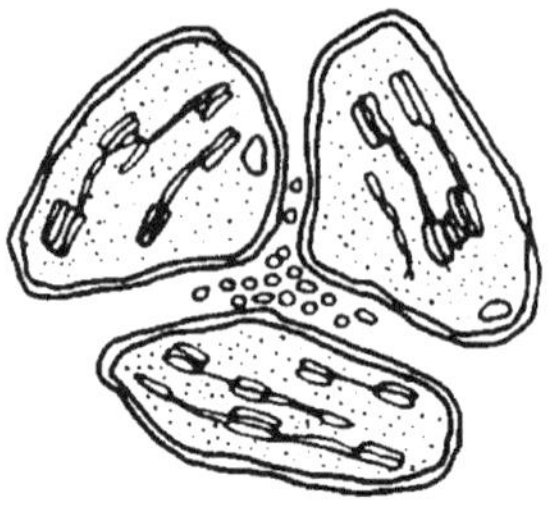
Tobacco mosaic virus in the chlorotic halo area of local lesion — Datura	Large osmiophilic globules accumulate in the chloroplast and the lamellae are distrupted.	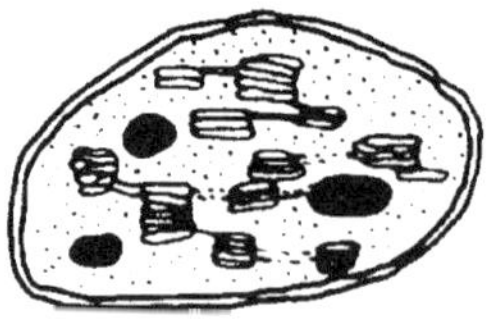

FIGURE 261

Structural abnormalities in chloroplasts associated with virus infections. (From Strobel, G.A. and Mathre, D.E., *Outlines of Plant Pathology,* 1st edition, by Van Nostrand Reinhold, New York, 1970, 282. © 1970. Reprinted with permission of Brooks/Cole, a division of Thomson Learning. Fax 800-730-2215.)

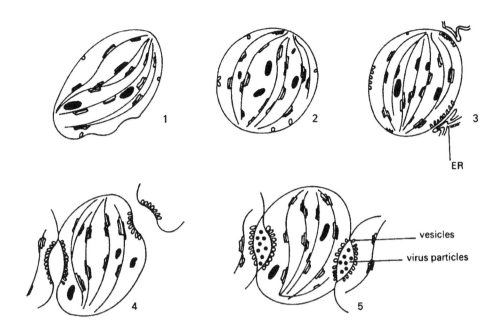

FIGURE 262

Sequence of changes in the chloroplasts of Chinese cabbage following infection with turnip yellow mosaic virus. (1) Chloroplast with scattered vesicles; (2) Swollen chloroplast, vesicles scattered. (3) Vesicles aggregated, endoplasmic reticulum (ER) associated with vesicles. (4) Chloroplasts aggregate together in area of vesicles. (5) Virus particles appear in space between chloroplasts. (After photomicrographs by Hatta, T., and Matthews, R.E.F., *Virology,* 59, 383, 1974; from Stevens, W.A., *Virology of Flowering Plants,* Blackie & Sons, Glasgow, 1983, 30. With permission of Kluwer Academic Publishers.)

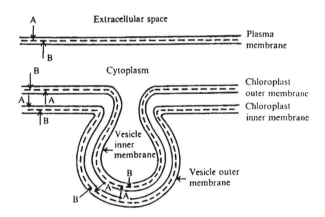

FIGURE 263

Peripheric membraneous vesicle induced by TYMV in chloroplasts. The diagram is based on the assumption that the vesicle membranes are derived from chloroplast membranes by invagination. A and B indicate membrane faces; fracture paths within the membranes are indicated by a dotted line. (From Hatta, T., Bullivant, S., and Matthews, R.E.F., *J. Gen. Virol.,* 20, 37, 1973. With permission.)

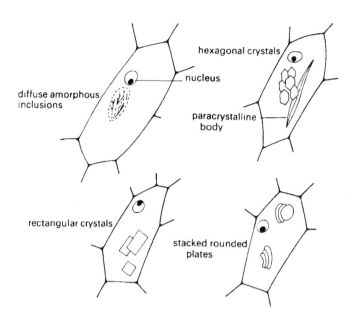

FIGURE 264
Various cytoplasmic inclusions induced by plant viruses. (From Stevens, W.A., *Virology of Flowering Plants,* Blackie & Sons, Glasgow, 1983, 35. With permission of Kluwer Academic Publishers.)

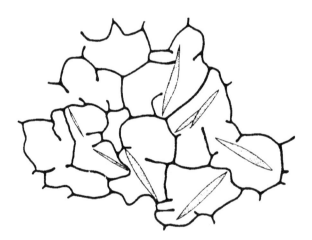

FIGURE 265
Fracture of intracellular membranes in a living plant tissue by sharp-pointed viral inclusions. The abundant growth of thread-shaped, needle-like, spindle-shaped virus inclusions (of a crystalloid or paracrystalloid kind) may hamper the normal course of cell division in virus-diseased plants. (From Goldin, M.I., in *Viruses of Plants,* Proc. Internat. Conf. Plant Viruses, Beemster, A.B.R. and Dijkstra, J., Eds., North-Holland Publishing Company, Amsterdam, 1966, 158. With permission of S. Chirkov, Russian Academy of Sciences, Moscow.)

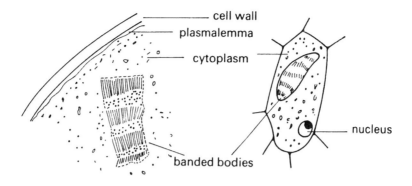

FIGURE 266

Cytoplasmic banded bodies in parenchyma cells. (From Stevens, W.A., *Virology of Flowering Plants,* Blackie & Sons, Glasgow, 1983, 36. With permission of Kluwer Academic Publishers.)

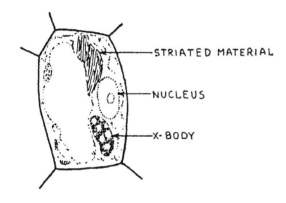

FIGURE 267

Cell with inclusion bodies characteristic of tobacco mosaic virus. The bodies are of two kinds: an amoeboid x-body and a body of crystalline striated material. The relative size of the bodies may be judged from their comparison with the nucleus. (From Esau, K., *Am. J. Bot.,* 43, 739, 1956. With permission.)

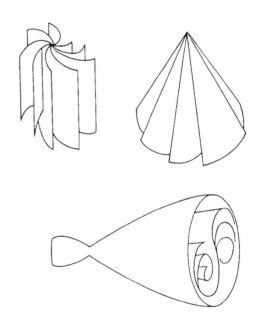

FIGURE 268

All members of the potyvirus family that have been investigated cytologically, induce the formation of intracellular cytoplasmic inclusions called "pinwheels." These formations, constituted by a type of protein encoded by the viral genome which assembles into regular sheets around an axis, were originally described as cylindrical. Some, at least during the initial stages of their development, may be conical in shape. (From Cornuet, P., *Éléments de Virologie Végétale,* © INRA, Paris. With permission.)

ABBREVIATIONS AND UNITS

ADCC	Antibody-dependent cell-mediated cytotoxicity
Ag	Antigen
AIDS	Acquired immunodeficiency syndrome
ALT	Alanine aminotransferase
APC	Antigen-presenting cell
BCAF	B-cell activation factor
BCDF	B-cell differentiation factor
BCGF	B-cell growth factor
BCl-2	A human gene whose function is related to apoptosis
BLV	Bovine leukemia virus
$\beta_2 M$	β_2 microglobulin
C, C'	Complement
CAEV	Caprine arthritis-encephalitis virus
CaMV	Cauliflower mosaic virus
CCC	Covalently closed circular DNA
CD	Cluster of differentiation of leucocyte antigen surface marker, e.g., CD4, CD8
CDC	Centers for Disease Control (Atlanta)
CF	Complement fixation
CMI	Cell-mediated immunity
CMV	Cytomegalovirus
CNS	Central nervous system
CPMV	Cowpea mosaic virus
CR	Complement receptor, e.g., CR3
CRP	C-reactive protein
CSF	Cerebrospinal fluid
CTL	Cytotoxic T lymphocyte
CTLp	CTL precursor
dpi	Dots per inch
DNA-p	DNA polymerase
dNTP	Deoxyribonucleoside triphosphate
ds	Double-stranded
DV	Dengue virus
EA/VCA	Early antigen/virus capsid antigen
EBV	Epstein-Barr virus
EBNA	Epstein-Barr virus-associated nuclear antigen
EGF	Epidermal growth factor
ELISA	Enzyme-linked-immuno-sorbent assay
ER	Endoplasmic reticulum
FADD	Fas-associating protein with death domain
FAIDS	Feline AIDS
Fas	A TNF-type receptor on the cell surface
FasL	Fas ligand
Fc	A fragment of IgG antibody
$Fc\gamma R$	Fc receptor for IgG
FCR	Fc receptor
FDC	Follicular dendritic cell
FeLV	Feline leukemia virus
FeSV	Feline sarcoma virus
FGF	Fibroblast growth factor
FLICE	A caspase
gp	Gene product, protein
GM	Granulocyte-macrophage
GM-CSF	Granulocyte-macrophage colony-stimulating factor
h, hr	hour
HA	Hemagglutinin, e.g., HA1, HA2

HAV	Hepatitis A virus
HBAg	Hepatitis B antigen
HBcAg	Hepatitis B core antigen
HBeAg	Hepatitis B e antigen
HBsAg	Hepatitis B surface antigen
HBV	Hepatitis B virus
HBx	HBx protein
HCC	Hepatocellular carcinoma
HCV	Hepatitis C virus
HDAg	Hepatitis D antigen
HDV	Hepatitis D (delta) virus
hIAP	Human inhibitor of apoptosis protein
HIV	Human immunodeficiency virus, e.g., HIV-1, HIV-2
HLA	Human leukocyte antigen (syn. MHC)
HPV	Human polyoma virus, e.g., HPV 16, HPV 18
Hsp	Human stress protein
HSV	Herpes simplex virus, e.g., HSV-1, HSV-2
HTLV	Human T-cell lymphotropic virus, e.g., HTLV-1
Ia	Antigen related to the MHI class pathway
i.c.	Intracerebral
ICAM	Intercellular adhesion molecule, e.g., ICAM-1
ICE	IL-1β converting enzyme
IEM	Immuno-electron microscopy
IF	Immunofluorescence
IFN	Interferon, e.g., IFN-α, IFN-β
Ig	Immunoglobulin, e.g., IgA, IgM, IgG
IL	Interleukin, e.g., IL-1, IL-2
IL-2R	IL-2 receptor
IRF	IFN-regulating factor
IRP	IFN-regulated protein
ISG	IFN-stimulated gene
ISGF	IFN-stimulated gene factor
ISRE	IFN-stimulated response element
i.v.	intravenous
IκB	Inhibitor of κB
K, KDa	Kilodalton
K cell	Kupffer cell, also killer cell
LAK	Lymphokine-associated killer cell
LCM	Lymphocytic choriomeningitis
LCMV	Lymphocytic choriomeningitis virus
LD50	Lethal dosis for 50% of subjects
LFA	Lymphocyte function-associated antigen, e.g., LFA-1, LFA-2
LMP	Latent membrane protein, e.g., LMP1, LMP2
LT	Lymphotoxin
M	Memory B cell
MAF	Macrophage-activating factor
M-CSF	Multi-colony-stimulating factor
MHC	Major histocompatibility complex (syn HLA), e.g., MHC-I, MHC-II
MIF	Migration inhibition factor
MMH	Mouse hepatitis virus
MMTV	Mouse mammary tumor virus
MOI, moi	Multiplicity of infection
mRNA	Messenger RNA
MxA	Mx protein A
MuLV	Murine leukemia virus
myc	Myelocytoma (viral oncoprotein)
Mϕ	Macrophage
μm	Micron, micrometer
NFκB	Nuclear (transcription) factor (binding to) κ locus of B lymphocytes
NK	Natural killer cell
NKCF	NK cytotoxic factor
NK-R	NK receptor

nm	Nanometer
NPV	Nuclear polyhedrosis virus
P	Plasma cell
p	Protein
PCH	Primary hepatic carcinoma
PCR	Polymerase-chain reaction
PFP	Pore-forming protein, perforin
PFU	Plaque-forming unit
PGE_2	Prostaglandin E_2
pIgA	Polymeric IgA
pIgR	Polymeric Ig receptor
PKC	Protein kinase C
PKR	dsRNA-activated protein kinase
PMN	Polymorphonuclear
PRD	Positive regulating domain
R	Receptor
Rb	Retinoblastoma
RSV	Respiratory syncytial virus
SMAF	Specific macrophage-arming factor
ss	Single-stranded
SSPE	Subacute sclerosing panencephalitis
TAP	Transport-associated protein
Tc	Cytotoxic T-cell
TCR	T-cell receptor
TfR	Transferrin receptor
TGF	Transforming growth factor, e.g., TGF-α
Th	Helper T-cell, e.g., Th1, Th2
TIP	Translation inhibitory protein
TMV	Tobacco mosaic virus
TNF	Tumor necrosis factor, e.g., TNF-α
TNF-R	TNF receptor
TRAF	TNF-R-associated factor
Ts	Suppressor T-cell
TYMV	Turnip yellow mosaic virus
T3-Ti	T-cell receptor
T=3	Capsid with triangulation number 3
vRNA	Viral RNA
(+)	Plus, positive-sense, message-sense, messenger-sense
(-)	Minus, negative-sense, anti-messenger sense

GLOSSARY

Abortive infection	Infection without production of infectious progeny
Acidophilic	Affinity for acidic dyes
Aggregation	Formation of a mass of particles
Aleutian mink disease	An economically important disease caused by a parvovirus
Alfalfa mosaic virus	Type species of *Alfamovirus* genus, *Bromoviridae* family
Amnion	Fetal membrane surrounding amniotic cavity and fetus (birds, reptiles, mammals)
Anergy	Functional inactivation of lymphocytes on encountering an antigen
Aneuploid	Abnormal (increased or decreased) chromosome number
Antibody	Immunoglobulin molecule produced in response to introduction of an antigen and able to combine specifically with that antigen
Antibody-dependent cellular cytotoxicity	Lysis of cells coated by a specific antibody by nonsensitized cells with Fc receptors able to recognize this antibody
Antigen	Substance able to induce a specific immune response
Antigenic tolerance	Absence of immune response after administration of antigen
Apoptosis	Programmed cell death
Arabis mosaic virus	Member of *Nepovirus* genus, *Comoviridae* family
Arbovirus	"Arthropod-borne", replicating in vertebrates and arthropods
Aseptic meningitis	Meningitis without bacteria in liquor
Assembly	Assembly of virus components, also called morphogenesis or maturation
Attachment	Adsorption, fixation
Aujeszky's disease	Pseudorabies, a herpesvirus disease of pigs and other animals
Autoimmunity	Immune reaction to self-antigens
Axon	Long threadlike outgrowth of nerve cells
B cell, B lymphocyte	Cell that mediates humoral immunity and is thymus-independent
Barley stripe mosaic virus	Type species of *Hordeivirus* genus
Basipetal	Movement from the apex to the base
Basophilic	Affinity for basic dyes
Beet curly-top virus	Type species of *Curtovirus* genus, *Geminiviridae* family
Biphasic	In two phases (steps)
Blast cell	Large immature cell of the hematopoietic system with abnormal morphology
Borna disease agent	An arbovirus causing chronic spasms and paralysis in horses, cattle, and other ungulates
Bowenoid	Resembling a precancerous dermatosis named after Bowen
Budding	Virus release from cellular membranes by formation of an outgrowth
Bundle sheath	Sheath around a vascular bundle consisting of a layer of parenchyma (BOT)
Burkitt's lymphoma	B cell lymphoma caused by Epstein-Barr virus, especially affecting children in Central Africa
Bursa of Fabricius	B cell-producing lymphoid organ in the hindgut of birds
Callus	Cover of hard tissue over a damaged plant surface (BOT)
Calyx or goblet cell	Mucus-secreting cell, e.g., in intestinal epithelium
Cambium	Cell layer between phloem and xylem responsible for increase in girth of stems and roots
Carcinoma *in situ*	Precursor stage of invasive carcinoma
Carrier	Generally asymptomatic, persistently infected individual
Caspase	ICE (IL-1β converting enzyme) protease
Cauliflower mosaic virus	Type species of *Caulimovirus* genus
Cellular immunity	Specific immunity mediated by T lymphocytes
Centripetal	Moving toward a center
Cervical	Pertaining to the neck or to the *cervix uteri*
Chemokine	Cytokine with chimiotactic activity toward nucleated blood cells
Chitins	Family of long-chain polymers of *N*-acetylglucosamine units
Chloroplast	Chlorophyll-containing organelle of photosynthetic plant cells
Ciliated	Provided with cilia (flagellae)
Circulative	Circulating in the hemocoel
Cistron	Basic unit of genetic function, usually a gene

Clearance	Elimination, removal
Clonal	Deriving from a single progenitor cell
Cocarcinogen	Secondary carcinogenic factor
Colony-stimulating factor	Cytokine that stimulates growth and differentiation of immature leukocytes in bone marrow
Complement	Set of normal nonspecific serum proteins that react in sequence and cause, among others, lysis of cells in the presence of antibodies
Conformational change	Change in three-dimensional arrangements
Contact inhibition	Tendency of normal cultured cells to cease growth upon contact with other cells
Cowdry type A	Intranuclear inclusion body in cells infected with human herpesviruses
Cowpea mosaic virus	Type species of *Comovirus* genus, *Comoviridae* family
Creutzfeldt-Jakob disease	Prion-caused spongiform encephalopathy
Cryptococcus	Capsulated yeast that causes systemic infections in AIDS and HTLV leukemia patients
Cured cell	Cell that has lost latent viruses or virus DNA
Cuticle	1. Layer of waxy material (cutin) on surface of many plants (BOT) 2. Exoskeleton of arthopods (INV)
Cutin	Mixture of interconnected polymeric fatty acids found on plant surfaces
Cytokine	Intercellular regulatory protein, generally produced by cells of the immune system, that acts on other cells
Cytolysis	Lysis of cells
Cytosol	Cytoplasm without organelles and cytoskeleton (eukaryotes)
Cytotoxic T-cell	T-cell able to kill other cells
Cytoxan	Cyclophosphamide, Endoxan; a mutagen and carcinogen
Dane particle	Hepatitis B virus
Defective virus	Incomplete virus unable to replicate
Dendritic cell	Dendrocyte; a particularly potent type of antigen-presenting cell
Dimer, dimeric	Consisting of two units
Ecdysis	Molting of the outer cuticular layer of the body (insects, arthropods)
Ectodesmata	Cytoplasmic extrusions through the outer wall of leaf epidermic cells
Effector	Cell (e.g., lymphocyte or phagocyte) that produces the end effect
Egestion	Expulsion of digestive secretions (INV)
egr	A retroviral oncogene
Egress	Exit of virus by budding or excretion
Enamovirus	Member and genus of *Luteoviridae* family
Endocytosis	Intake via membrane-bound vesicles
Endonuclease	Enzyme that cleaves phosphodiester bonds within DNA or RNA
Endoplasmic reticulum	Intracellular system of channels and vesicles (cisternae) contiguous with nuclear membrane (eukaryotes)
Endosome	Organelle of the endocytic pathway, located between plasma membrane and lysosomes
Entomopoxvirus	Insect poxvirus
Envelope	Lipoprotein membrane surrounding a capsid or nucleoprotein (VIR)
Epicuticle	Outer waxy layer of the exoskeleton of many arthropods
Episome	Genetic element able to replicate independently or as integrated part of a cellular genome (see plasmid)
Epizootic	Epidemic disease among animals
Epstein-Barr virus	Gammaherpesvirus discovered by Epstein and Barr in 1964
Erythropoietic	Forming red blood cells
Erythropoietin	Glycoprotein hormone that regulates the formation of red blood cells
Fas	Receptor that signals programmed cell death
Fc	IgG immunoglobulin fragment produced by papain digestion
Fc receptor	Cell surface protein able to bind the Fc fragment
Fiji disease virus	Type species of *Fijivirus* genus, family *Reoviridae*
Follicle	Region in lymphoid tissue that contains mostly B cells
fos	A retroviral oncogene
Foscarnet	Antiviral, synonym phosphonoformate
fra	A retroviral oncogene
Friend tumor virus	Member of genus "Mammalian C type virus", family *Retroviridae*

G1 phase	Gap in cell cycle after mitosis and before DNA replication
G0 phase	Subdivision of G1, pause in cell growth
Gall	Abnormal outgrowth from plant stem or leaf caused by parasite invasion (BOT)
Gamete	Germ cell able to fuse with another in sexual reproduction
Ganciclovir	Antiviral, a guanosine analog
Genomic	Relating to the genome
Germ line	Series of cells showing continuous genetic information between generations
Glycoprotein	Protein with covalently linked carbohydrate moiety
Glycosylation	Biosynthesis of glycoproteins
Golgi apparatus	Stack of flattened vesicles forming an intracellular transport system in eukaryotes
Grana	Stack of multilayered membranes within chloroplast
Granulin	Protein case surrounding a granulosis virion
Granulosis	Lepidopteran disease caused by granulosis baculoviruses
Granzyme	Fragmentine; serin esterase that induces apoptosis by fragmenting cellular DNA
Grasserie	Silkworm disease caused by a nuclear polyhedrosis baculovirus
Guarnieri body	Cytoplasmic inclusion body in cells infected with vaccinia and related viruses
HBx protein	Small HBV protein with *trans*-activating function
Hemocoel	Blood-filled cavity consisting of empty spaces between organs (INV)
Hemolymph	Fluid (blood equivalent) in open circulatory systems of many invertebrates
Hindgut	Terminal part of bird or invertebrate intestines
Histamine	Histidine-derived basic amine with many physiologic effects
Homeostasis	Maintenance of *status quo*
Humoral immunity	Immune phenomena implicating B cells and specific antibodies in a body fluid
Hyperplasia	Enlargement of an organ or tissue due to increased cell number
Hypertrophia	Enlargement of an organ or tissue due to increased cell size
Hypoplasia	Deficient development
Iatrogenic	Caused by medical examination or treatment
Inapparent infection	Infection without obvious symptoms
Inclusion body	Microscopically visible intracellular aggregate of viruses and/or protein
Integrase	Virus-mediated enzyme which mediates integration
Integration	Insertion of viral DNA into the host genome
Integrins	Dimeric cell membrane glycoproteins that serve as extracellular receptors
Integument	Outer covering or coating
Interferon	Cytokine of stimulated vertebrate cells that interferes with virus infection, cell growth, and the immune response
Interferon type I	Interferons of types α and β
Interleukin	Cytokine produced by leukocytes that acts on other leukocytes
Isotropic	Having a tendency for equal growth in all directions
JC virus	Polyomavirus associated with progressive multifocal leucoencephalopathy
jun	A retroviral oncogene
Junín virus	Agent of Argentinian hemorrhagic fever
Kaposi's sarcoma	Angioproliferative, often multifocal tumor involving skin, mucosa, and viscera
Killer T cell	Cytotoxic T cell
Koilocyte	Cell of the uterine cervix with abnormal nuclei surrounded by transparent cytoplasm
Koplik's spots	White spots on oral mucosa in the first stage of measles
Kupffer cell	Mononuclear fixed phagocytic cell lining liver sinuoids
Lamella	Thin scale or plate; photosynthetic in Fig. 261
Lamina	Thin sheet or layer of tissue
Langerhans cell	Type of antigen-presenting cell in skin and parts of the intestine
Lansing virus	Prototype strain of human poliovirus II
Lassa fever	Human hemorrhagic fever caused by an arenavirus
Latency	Stage of virus infection excluding the symptom-less incubation period
Latent virus	Virus present without clinical manifestations
Lepidoptera	Order of insects with two pairs of scaly wings in the adult stage; comprises butterflies and moths
Lieberkühn's crypts	Pit-like glands at the basis of intestinal villi
Ligand	Substance able to bind

Luteovirus	Member and genus of *Luteoviridae* family
Lymphocyte	Type of white blood cell which is involved in the immune response
Lymphokine	Cytokine from activated lymphocytes that acts on cells of the immune system
Lymphoreticular tissue	Sum of lymphoid and reticularendothelial tissues
Lysis	Disruption of (host) cell
Lysosome	Intracytoplasmic membraneous vesicle containing hydrolytic enzymes in acidic pH
Lysozyme	Enzyme which hydrolyzes 1,4-β glycosidic bonds in mucopolysaccharides and mucopeptides in bacterial cell walls and chitin
M cell	Mononuclear cell (in this context)
Macrophage	Large mononuclear phagocyte found in tissue (histiocyte) and blood (monocyte), necessary for antigen-processing
Maedi-Visna virus	Icelandic, *maedi*, "labored breathing"; *visna*, "paralysis and wasting"; member of *Lentivirus* genus, family *Retroviridae*
Major histocompatibility complex (MHC)	A large region of cell DNA containing genes for histocompatibility antigens whose function is antigen-processing and presentation
Marburg virus	Type species of *Filovirus* genus, family *Filoviridae*
Marek's disease	Infection of fowl causing progressive paralysis and lymphoid tumors
Mast cell	Tissue equivalent of basophil blood cells, contains serotonin and histamine granules
Mediator	Agent which brings about an effect
Medulla	Central region of lymph node
Membrane fusion	1. Fusion of cytoplasmic membranes of cells
	2. Fusion of cell membrane and viral envelope
Memory B cell	Long-lived B cell with memory (primed) for an antigen
Mesophyll	Parenchymatous tissue between upper and lower epidermal layer of leaves
Midgut	Middle portion of invertebrate intestine
Migration inhibitory factor	Lymphokine from sensitized lymphocytes that inhibits macrophage migration
Mitogen	Substance that stimulates lymphocytes to proliferate
Modulation	Regulation, adjustment
Monocyte	Large mononuclear cell in the blood
Multiplicity of infection	Ratio of infectious viruses to host cells
Mutagenic	Causing a mutation
Mx proteins	Antiviral GTP-binding proteins controlling, among others, influenza pathogenicity in mice
myc	A retroviral oncogene
Myocyte	Muscle cell
Myxoma	Tumor of mucous or gelatinous tissue
Naive T cell	T cell before an encounter with an antigen-presenting cell
"Natural killer" cell	Lymphocyte that kills antibody-coated, tumor-, and virus-infected cells
Negative-sense	(-) sense; nucleic acid is complementary to the (+) strand of viral mRNA
Negri body	Cytoplasmic inclusion body in cells infected with rabies virus
Neutralization	Inactivation of virus infectivity by specific antibody
NFκβ	Transcription enhancer binding to a sequence in the κlocus of Ig genes in B lymphocytes
Nodule formation	Aggregation of foreign elements by hemocyte contents and subsequent encapsulation by plasmocytes (INV)
Noncirculative	Not circulating in the hemocoel
Nonpropagative	Without virus propagation
Norwalk agent	Member of *Caliciviridae* family
Nucleocapsid	Nucleic acid + capsid
Nucleosome	Histone-containing repeating subunit in eukaryotic chromatin
Occlusion body	Large protein crystal which includes (occludes) viral particles
Oligoclonal	Cells derived from a few progenitor cells
Oncogene	Gene that promotes tumor development
Oncogenesis	Generation and development of tumors
Oncogenic	Causing tumor development
Opsonization	Coating with antibodies or complement factors (CB3) to facilitate phagocytosis
Orf virus	Type species of *Parapoxvirus* genus, family *Poxviridae*
Osmiophilic	Stining with osmic acid
Oviduct	Tube that conducts ova from the ovary
Oviposition	Laying eggs (insects)
Ovipositor	Tubular organ with which a female insect deposits her eggs

Pantropic	Affecting every system or organ
Paracrine	Acting closely to its place of origin
Parenchyma	1. Functional part of an organ as distinguished from connective and supporting tissue (AN)
	2. Plant tissue consisting of cells that are agents of photosynthesis and storage
-partite	Consisting of parts
Pectins	Group of complex polysaccharides consisting chiefly of galacturonic acid
Peplomer	Glycoprotein projection of viral envelope, also called "spike"
Perforin	Pore-forming protein of NK cells, forms tubular lesions on target cell membranes
Peritrophic membrane	Membrane fold in the midgut of insects
Permissive	Able to support a productive virus infection
Persistent infection	Equilibrium host-virus without lysis
Peyer's patches	Areas of lymphoid tissue in the submucosa of the small intestine
Phagocytose	Engulfment of particles and viruses into the cytoplasm
Phloem	Nutrient-conducting vascular tissue in higher plants
Plasma cell	Cell of the B lymphocyte lineage, main producer of immunoglobulins
Plasmalemma	Cytoplasmic membrane
Plasma membrane	Cytoplasmic membrane
Plasmid	Extrachromosomal genetic element, generally circular dsDNA
Plasm(at)ocyte	Amoeboid blood cells with phagocytic ability (INV)
Plasmodesmata	Intercellular cytoplasmic bridges between adjacent cells (BOT)
Pneumocystis carinii	A fungus that causes pneumonia in AIDS and HTLV leukemia patients
Poly(A) tract	Polymer of adenosine nucleotides
Polyhedron	Solid body with many faces
Polyhedrosis	Invertebrate (mainly insect) disease characterized by the presence of nuclear or cytoplasmic inclusion bodies
Polymerase	Enzyme catalyzing DNA or RNA synthesis
Polysome	Polyribosome
Potyvirus	Genus name, also member of *Potyviridae* family
Prion	"Proteinaceous infectious particle"
Productive infection	Infection with production of progeny virus
Progeny	Daughter generation in virus replication
Proinflammatory	Inducing an inflammatory response
Promoter	Part of gene which controls initiation and rate of transcription
Properdin	Protein (factor P) involved into the alternative complement pathway
Prostaglandin	Type of local chemokines that stimulate inflammation
Proteasome	Cellular organelle involved into the class I MHC pathway
Provirus	1. dsDNA copy of an ssRNA retroviral genome
	2. Any latent (integrated) virus genome
	3. Immature virus (obsolete)
Quinolinic acid	Precursor of nicotinic acid
Read-through protein	Protein resulting from the reading of a mRNA through a stop codon
Receptor	Virus-binding site on cell surface
Replicative intermediate	Complex of newly synthetized daughter RNA strands and their template
Restrictive infection	Viral replication starts, but is blocked at an early stade
Reticulo-endothelial system	Sum of macrophages, endothelial cells, and polymorphonuclear leukocytes
Retinoblastoma protein	A phosphoprotein with DNA-binding activity
Retrotransposon	Cellular DNA element which transposes (inserts itself) via a RNA intermediate
Reverse transcriptase	RNA-dependent DNA polymerase for DNA synthesis on RNA templates
Ribosome	Ribonucleoprotein particle which is the site of protein synthesis
S phase	Part of the cell cycle (DNA synthesis phase)
Sandfly fever	Recurrent fever caused by a bunyavirus of genus *Phlebovirus*
Schwann cell	Glial cell which forms the fatty sheath around myelinated nerve fibers
Scrapie	Transmissible spongiform encephalopathy of sheep
Shedding	Dispersal into the environment
Shope papilloma	Infection of cotton-tail rabbits caused by a papillomavirus
Sialic acid	N-acetylneuraminic acid, a constituent of many glycoproteins
Sieve cell or element	Food-containing, long, tapering phloem cell of gymnosperms and lower vascular plants
Sindbis virus	Type species of *Alphavirus* genus, family *Togaviridae*

Sink leaf	Immature leaf
Slow virus	Virus causing slow progressive disease of the CNS
Somatic cell	Body cell (excluding gametes or their precursors)
Southern bean mosic virus	Type species of *Sobemovirus* genus
Stem cell	Undifferentiated founder cell of embryonic or other lineage
Stem line	Lineage of stem cells
Superantigen	Antigen able to stimulate multiple T-lymphocytes and causing a massive and harmful cell-mediated immune response
Superhelical	State of dsDNA in which the double helix is further twisted, also "supercoiled"
Symplast	Three-dimensional system of intercellular channels in plants
Syncytium	Multinucleated giant cell resulting from cell fusion
T-cell, T lymphocyte	Cell that mediates cellular immunity, is thymus-dependent, and does not produce antibodies
5'-Terminal cap	Sequence of methylated bases at the 5' end of certain viral RNAs
Ternary	Composed of three parts
T helper cell	Cell that stimulates B cells to develop into plasma cells and produce antibodies
Th2 cell	$CD4^+$ T-cell subset
Thylacoid	Membrane vesicle and chloroplasts
Tobacco mosaic virus	Type species of *Tobamovirus* genus
Trans-activation	Exercise of an influence far away
Transcription	mRNA synthesis from a DNA or RNA template
Transcytosis	Transfer across epithelial and endothelial cells by means of membrane-bound vesicles
Transformation	1. Gene transfer and expression of by intake of "naked" exogenous DNA 2. Induction of a tumor-like lifestyle in a cell
Translation	Protein synthesis directed by RNA
Translocation	1. Movement of a polypeptide across a biological membrane (IMM) 2. Movement of sugars through the phloem in vascular plants (BOT)
Triangulation number	Number of equilateral triangles contained on the face of an icosahedron
Trophoblast	Early embryonic stage before embedding into the uterus
Tumor necrosis factor	Cytokine toxic for tumors, with antiviral activity
Turnip yellow mosaic virus	Type species of *Tymovirus* genus
Ubiquitin pathway	Pathway for intracellular protein degradation involving the protein ubiquitin
Uncoating	Removal of outer layers of viral particles and exposure of viral nucleic acids
Unwindase	Protein that unwinds origins of replication in DNA
Uukuniemi virus	Member of *Phlebovirus* genus, family *Bunyaviridae*
Vegetative replication	Replication with production of progeny virus
Vertical transmission	Transmission from parent to progeny
Viremia	Presence and spread of viruses in the bloodstream
Virion	Complete infectious viral particle
Virogenic stroma	Amorphous, membrane-less inclusion body associated with virus production
Virolysis	Lysis by perforation of viral envelopes by antibodies or complement
Virus factory	Amorphous membrane-less inclusion body associated with virus production
Xylem	Principal water-conducting and supporting (woody) tissue of higher plants
Zoonosis	Disease which is naturally transmitted to man by animals
Zoospore	Motile flagellated spore
AN	In anatomy
BOT	In botany
IMM	In immunology
INV	In invertebrates
VIR	In virology

REFERENCES

1. **Ackermann, H.-W. and Berthiaume, L.,** *Atlas of Virus Diagrams,* CRC Press, Boca Raton, 1995.
2. **Ackermann, H.-W., Berthiaume, L., and Tremblay, M.,** *Virus Life in Diagrams,* CRC Press, Boca Raton, 1998.
3. **Cruse, J.M. and Lewis, R.E.,** *Illustrated Dictionary of Immunology,* CRC Press, Boca Raton, 1995.
4. **Hull, R., Brown, F., and Payne, C.,** *Virology, Directory and Dictionary of Animal, Bacterial, and Plant Viruses,* Macmillan Press, London, 1989.
5. **Kendrew, J.,** Ed.-in-chief, *The Encyclopedia of Molecular Biology,* Blackwell Science, Oxford, 1994.
6. **Parker, S.P.,** Ed.-in-chief, *McGraw-Hill Dictionary of Biology,* 3rd ed., McGraw-Hill, New York, 1984.
7. **Matthews, R.E.F.,** A history of viral taxonomy, in *A Critical Appraisal of Viral Taxonomy,* Matthews, R.E.F., Ed., CRC Press, Boca Raton, 1983, 1.
8. **Van Regenmortel, M.H.V., Fauquet, C.M., Bishop, D.H.L., Carstens, E., Estes, M.K., Lemon, S., Maniloff, J., Mayo, M.A., McGeoch, D.J., Pringle, C.R., and Wickner, R.,** Eds., *Virus Taxonomy. Classification and Nomenclature of Viruses. Seventh Report of the International Committee on Taxonomy of Viruses*, Academic Press, New York, in print.
9. **Lwoff, A., Horne, R.W., and Tournier, P.,** A system of viruses, *Cold Spring Harbor Symp. Quant. Biol.,* 27, 51, 1962.
10. **Pringle, C.R.,** Virus taxonomy - 1999, *Arch. Virol.,* 144, 421, 1999.
11. **Maniloff, J. and Ackermann, H.-W.,** Taxonomy of bacterial virus genera and the order *Caudovirales, Arch. Virol.,* 143, 2051, 1998.
12. **Ward, C.W.,** Progress toward a higher taxonomy of viruses, *Res. Virol.,* 144, 419, 1993.
13. **Van Regenmortel, M.H.V.,** Virus species, a much overlooked but essential concept in virus classification, *Intervirology,* 31, 241, 1990.
14. **Kozak, M.,** Regulation of translation in eukaryotic systems, *Annu. Rev. Cell Biol.,* 8, 197, 1992.
15. **Meinl, E., Fickenscher, H., Thome, M., Tschopp, J., and Fleckenstein, B.,** Anti-apoptotic strategies of lymphotropic viruses, *Immunol. Today,* 19, 474, 1998.
16. **Isaacs, A., and Lindenmann, J.,** Virus interference. I. The interferon, *Proc. Roy. Soc. London Ser. B.,* 147, 258, 1957.
17. **Tremblay, M.J., Fortin, J.-F., and Cantin, R.,** The acquisition of host-encoded proteins by nascent HIV-1, *Immunol. Today,* 19, 346, 1998.
18. **Evans, A.S.,** Epidemiological concepts and methods, in *Viral Infections of Humans,* 3rd ed., Evans, A.S., Ed., Plenum Press, New York, 1982, 3.
19. **Mims, C.A. and White, D.O.,** *Viral Pathogenesis and Immunology,* Blackwell Scientific Publications, Oxford, 1984, 40, 50, 52, 55, 61, 72, 90, 208, and 246.
20. **Stringfellow, D.A.,** Viral pathogenesis and host resistance to infection, in *Virology,* Stringfellow, D.A., Ed.-in-chief, Upjohn, Kalamazoo, 1983, 45.
21. **Fenner, F., McAuslan, B.R., Mims, C.A., Sambrook, J., and White, D.O.,** *The Biology of Animal Viruses,* 2nd ed., Academic Press, New York, 1974, 342, 347, 384, and 453.
22. **Johnson, R.T., Narayan, O., and Clements, J.,** Varied role of viruses in chronic neurologic diseases, in *Persistent Viruses,* Stevens, J.G, Todaro, G.J., and Fox, C.F., Eds., Academic Press, New York, 1978, 551.
23. **Maréchal, V. , Dehée, A., and Nicolas, J.-C.,** Marqueurs de réplication et physiopathologie des infections virales, *Virologie,* 1 (special issue), 11, 1997.
24. **Bellanti, J.A.,** *Immunology,* W.B. Saunders, Philadelphia, 1971, 17, 64, 67, 271, and 286.
25. **Robinson, T.W.E. and Heath, R.B.,** *Virus Diseases and the Skin,* Churchill Livingstone, Edinburgh, 1983, 23.
26. **Anonymous,** *Viral and Rickettsial Diseases, Physician's Handbook,* 4th ed. Ontario Department of Health, Toronto, 1972, 29.
27. **Huraux, J.M., Nicolas, J.C., and Agut, H.,** in *Virologie,* Flammarion Médecine Sciences, Paris, 1985, 21 and 23.
28. **Johnson, R.T.,** Pathogenesis of viral infections, in *Viral Infections in Oral Medicine,* Hooks, J. and Jordan, G., Eds., Elsevier/North-Holland, New York, 1982, 3.
29. **Fenner, F.,** *The Biology of Animal Viruses,* Vol. 2, *The Pathogenesis and Ecology of Viral Infections,* Academic Press, New York, 1969, 517.
30. **Bale, J.F. and Kern, E.R.,** Viral infections of the nervous system, in *Virology,* Stringfellow, D.A., Ed., Upjohn, Kalamazoo, 1983, 77.
31. **Johnson, R.T.,** *Viral Infections of the Nervous System*, Raven Press, New York, 1982, 44, 47, 134, 137, 205, 298, and 299. Second edition published in 1995 by Lippincott-Raven, Philadelphia.
32. **Cottrall, G.E.,** Endogenous viruses in the egg, *Ann. N.Y. Acad. Sci.,* 55, 221, 1952.
33. **Notkins, A.I.,** Viral infections: mechanisms of immunologic defense and injury, *Hosp. Pract.,* 9, 65, 1974.
34. **Jindal, S. and Malkowsky, M.,** Stress responses to viral infection, *Trends Microbiol,* 2, 89, 1994.

35. **Kerr, J.F.R. and Harmon, B.V.,** Definition and incidence of apoptosis: an historical perspective, in *Apoptosis: The Molecular Basis of Cell Death. Current Communications in Cell and Molecular Biology,* Vol. 3, Tomei, L.D. and Cope, F.O., Eds., Cold Spring Harbor Laboratory Press, Cold Spring Harbor, 1991, 5.

36. **Gillet, G. and Brun, G.,** Viral inhibition of apoptosis, *Trends Microbiol.,* 4, 312, 1996.

37. **Griffin, D.E. and Hardwick, J.M.,** Regulators of apoptosis on the road to persistent alphavirus infection, *Annu. Rev. Microbiol.,* 51, 565, 1997.

38. **Roitt, I.M.,** *Essential Immunology,* 4th ed., Blackwell Scientific Publications, Oxford, 1980, 83, 84, and 215.

39. **Blacklaws, B., Bird, P., and McConnell, I.,** Early events in infection of lymphoid tissue by a lentivirus, maedi-visna, *Trends Microbiol.,* 3, 434, 1995.

40. **Baron, S. and Dianzini, F.,** General considerations on the interferon system, in *The Interferon System: A Current Review to 1978,* Baron, S. and Dianzini, F., Eds., *Texas Rep. Biol. Med.,* 35, 1, 1977.

41. **Fauci, A.S., Rosenberg, S.A., Sherwin, S.A., Dinarello, C.A., Longo, D.L., and Lane, H.,** Immunomoderators in clinical medicine, *Ann. Int. Med.,* 106, 421, 1987.

42. **Sen, G.C. and Ransohoff, R.M.,** Interferon-induced antiviral actions and their regulation, *Adv. Virus Res.,* 42, 57, 1993.

43. **Marcus, P.I. and Salb, J.M.,** Molecular basis of interferon action: inhibition of viral RNA translation. *Virology,* 30, 502, 1966.

44. **Samuel, C.E.,** Interferon, in *Encyclopedia of Virology,* Vol. 2, Lederberg, J., Ed.-in-chief, Academic Press, San Diego, 1992, 533.

45. **Levine, A.J.,** *Viruses,* Scientific American Library, New York, 1992, 55.

46. **Norley, S. and Kurth, R.,** Immune response to retroviral infection, in *The Retroviridae,* Vol. 3., Levy, J.A., Ed., Plenum Press, New York, 1992, 363.

47. **Maletic Neuzil, K. and Graham, B.S.,** Cytokine release and innate immunity in respiratory virus infection, *Sem. Virol.,* 7, 255, 1996.

48. **Smith, T.J., Mosser, A.G., and Baker, T.S.,** Structural studies on the mechanisms of antibody-mediated neutralization of human rhinovirus, *Sem. Virol.,* 6, 233, 1995.

49. **Lamm, M.E.,** Interactions of antigens and antibodies at mucosal surfaces, *Annu. Rev. Microbiol.,* 51, 311, 1997.

50. **Slifka, M.K. and Ahmed, R.,** Long-term immunity against viruses: revisiting the issue of plasma cell longevity, *Trends Microbiol.,* 4, 394, 1996.

51. **Alexander, J.W. and Good, R.A.,** *Fundamentals of Clinical Immunology,* W.B. Saunders, Philadelphia, 1977, 103.

52. **Unanue, E.R. and Benacerraf, B.,** *Textbook of Immunology,* 2nd ed., Williams & Wilkins, Baltimore, 1984, 162.

53. **Sanfilippo, A., Balber, A.E., Granger, D.I., and McKinney, R.E.,** Immune responses to infection, in *Zinsser Microbiology,* 19th ed., Joklik, W.K., Willett, H.P., Amos, D.B., and Wilfert, C.M., Eds., Appleton & Lange, Norwalk, 1988, 277.

54. **Trinchieri, G.,** Biology of natural killer cells, *Adv. Immunol.,* 47, 187, 1989.

55. **Whitton, J.L. and Oldstone, M.B.A.,** Immune response to viruses, in *Fields Virology,* 3rd ed., Vol. 1, Fields, B.N., Knipe, D.M., and Howley, P.M., Eds.-in-chief, Lippincott-Raven, Philadelphia, 1996, 345.

56. **Hilleman, M.R.,** Historical and contemporary perspectives in vaccine development: from the vantage of cancer, *Progr. Med. Virol.,* 39, 1, 1992.

57. **Schüpbach, J.,** *Human Retrovirology, Facts and Concepts, Curr. Topics Microbiol. Immunol.,* 142, Springer-Verlag, Berlin, 1989, 36 and 52.

58. **Burgert, H.-G.,** Subversion of the MHC class I antigen-presentation pathway by adenoviruses and herpes simplex virus, *Trends Microbiol.,* 4, 107, 1996.

59. **Brutkiewicz, R.R. and Welsh, R.M.,** Major histocompatibility complex class I antigens and the control of viral infections by natural killer cells, *J. Virol.,* 69, 3967, 1995.

60. **Doherty, P.C.,** Immune exhaustion: driving virus-specific CD8$^+$ T cells to death. *Trends Microbiol.,* 1, 207, 1993.

61. **Lee, S.P.,** Control mechanisms of Epstein-Barr virus persistnce, *Sem. Virol.,* 5, 281, 1994.

62. **Von Herrath, M.G. and Oldstone, M.B.A.,** Role of viruses in the loss of tolerance to self-antigens and in auto-immune diseases, *Trends Microbiol.,* 3, 424, 1995.

63. **White, E.,** Function of the adenovirus E1B oncogene in infected and transformed cells, *Sem. Virol.,* 5, 341, 1994.

64. **Bausch, D.G. and Rollin, P.E.,** La fièvre de Lassa, *Ann. Inst. Pasteur/Actualités,* 8, 223, 1997.

65. **Farmer, T.W. and Janeway, C.A.,** Infections with the virus of lymphocytic choriomeningitis, *Medicine,* 21, 1, 1942.

66. **Oldstone, M.B.A.,** Immunotherapy for virus infection, in *Arenaviruses. Biology and Immunotherapy,* Oldstone, M.B.A., Ed., *Curr. Topics Microbiol. Immunol.,* 134, Springer, Berlin, 1987, 211.

67. **Weissenbacher, M.C., Laguens, R.P., and Coto, E.E.,** Argentine hemorrhagic fever, in *Arenaviruses. Biology and Immunotherapy,* Oldstone, M.B.A., Ed., *Curr. Topics Microbiol. Immunol.,* 134, Springer, Berlin, 1987, 79.

68. **Parsonson, I.M. and McPhee, D.A.,** Bunyavirus pathogenesis, *Adv. Virus Res.,* 30, 279, 1985.

69. **Kozuch, O. Rajcáni, J., Sekeyová, M., and Nosek, J.,** Uukuniemi virus in small rodents, *Acta Virol.,* 14, 163, 1970.

70. **Purcell, R.H.,** Hepatitis E virus, in *Fields Virology,* 3rd ed., Vol. 2, Fields, B.N., Knipe, D.M., and Howley, P.M., Eds.-in-chief, Lippincott-Raven, Philadelphia, 1996, 2831.

71. **Perlman, S., Jacobsen, G., Olson, A.L., and Afifi, A.,** Identification of the spinal cord as a major site of persistence during chronic infection with a murine coronavirus, *Virology,* 175, 418, 1990.

72. **Pedersen, N.C., Black, J.W., Boyle, J.F., Evermann, J.F., McKeirnan, A.J., and Ott, R.L.,** Pathogenic differences between feline coronavirus isolates, in *Molecular Biology and Pathogenesis of Coronaviruses,* Rottier, P.J.M., Van der Zeijst, B.A.M., Spaan, W.J.M., and Horzinek, M.C., Eds., Plenum Press, New York, 1983, 365.

73. **Andries, K. and Pensaert, M.,** Vomiting and wasting disease, a coronavirus infection of pigs, in *Biochemistry and Biology of Coronaviruses,* Ter Meulen, V., Siddell, S., and Wege, H., Eds., Plenum Press, New York, 1981, 399.

74. **Feldmann, H., Volchkov, V.E., and Klenk, H.-D.,** Filovirus Ebola and Marburg, *Ann. Inst. Pasteur/Actualités,* 8, 207, 1997.

75. **Martini, G.A.,** Marburg virus disease, in *Marburg Virus Disease,* Martini, G.A. and Siegert, R., Eds., Springer, New York, 1971, 1.

76. **Stille, W. and Böhle, E.,** Clinical course and prognosis of Marburg ("green-monkey") disease, in *Marburg Virus Disease,* Martini, G.A. and Siegert, R., Eds., Springer, New York, 1971, 10.

77. **Nathanson, N.,** Pathogenesis, in *St. Louis Encephalitis,* Monath, T.P., Ed., American Public Health Association, Washington, 1980, 201.

78. **Monath, T.P.,** Central nervous system infections (acute), in *Virology and Rickettsiology,* Vol. I, Part 2, Hsiung, G.-D and Green, R.H., Eds., CRC Press, Boca Raton, 1978, 261.

79. **Kerr, J.A.,** Yellow fever as a model for the study of arthropod-borne virus infections. *Indian J. Med. Sci.,* 7, 338, 1953.

80. **Bielefeldt-Ohmann, H.,** Pathogenesis of dengue virus diseases: Missing pieces in the jigsaw, *Trends Microbiol.,* 5, 409, 1997.

81. **Kurane, L. and Ennis, F.A.,** Cytokines in dengue virus infections: Role of cytokines in the pathogenesis of dengue hemorrhagic fever, *Sem. Virol.,* 5, 443, 1994.

82. **Shimotohno, K.,** Hepatocellular carcinoma in Japan and its linkage to infection with hepatitis C virus, *Sem. Virol.,* 4, 305, 1993.

83. **Sherlock, S.,** *Diseases of the Liver and Biliary System,* 6th ed., Blackwell Scientific Publications, Oxford, 1981, 256 and 271.

84. **Garrigue, G.,** Hépatites virales, in *Virologie Médicale,* Maurin, J., Ed., Flammarion Médecine Sciences, Paris, 1985, 752.

85. **Vasseur, M.,** *Les Virus Oncogènes,* Hermann Éditeurs, Paris, 1989, 232.

86. **Hoofnagle, J.H. and Di Bisceglie, A.M.,** Antiviral therapy of hepatitis, in *Antiviral Agents of Viral Diseases of Man,* 3rd ed., Galasso, G.J., Whitley, R.J., and Merigan, T.C., Eds., Raven Press, New York, 1990, 415.

87. **Eddleston, A.L.W.F.,** Acute hepatitis and fulminant hepatic failure, in *Oxford Textbook of Medicine,* 2nd ed. Vol. 1, Weatherall, D.J., Ledingham, L.G.G., and Warrell, D.A., Eds., Oxford University Press, Oxford, 1987, 12.214 and 12.216.

88. **Zoulim, F. and Trepo, C.,** Virus de l'hépatite B: réplication et mécanismes d'action des antiviraux, *Virologie,* 1, 197, 1997.

89. **Bianchi, L. and Gudat, F.,** Immunopathology of chronic hepatitis, in *Chronic Hepatitis,* Liaw, Y.-F., Ed., Excerpta Medica, Amsterdam, 1985, 45.

90. **Zoulim, F.,** Therapy of chronic hepatitis B virus infection: inhibiton of the viral polymerase and other antiviral strategies, *Antiviral Res.,* 44, 1, 1999.

91. **Thomas, H.C.,** Management of chronic hepatitis virus infections, in *New Antiviral Strategies,* Norrby, S.R., Mills, J., Norrby, E., and Whitton, L.J., Eds., Churchill Livingstone, Edinburgh, 1988, 52.

92. **Butel, J.S., Lee, T.-H., and Slagle, B.L.,** Is the DNA repair system involved in hepatitis-B-mediated hepatocellular carcinogenesis, *Trends Microbiol.,* 4, 119, 1996.

93. **Shafritz, D.A. and Hadziyannis, S.J.,** Hepatitis B virus DNA in liver and serum, viral antigens and antibodies, virus replication, and liver disease activity in patients with persistent hepatitis B virus infection, in *Advances in Hepatitis Research,* Chisari, F.V., Ed., Masson Publishing USA, New York, 1984, 241.

94. **Blumberg, B.S., Millman, I., Venkateswaran, P.S., and Thyagarajan, S.P.,** Hepatitis B virus and hepatocellular carcinoma - treatment of HBV carriers with *Phyllanthus amarus, Cancer Detect. Prev.,* 14, 195, 1989.

95. **Fu, X.-X., Su, C.Y., Lee, Y., Hintz, R., Biempica, L., Snyder, R., and Rogler,** Insulinlike growth factor II expression and oval cell proliferation associated with hepatocarcinogenesis in woodchuck hepatitis virus carriers, *J. Virol.,* 62, 3422, 1988.

96. **Purcell, R.H., Hoofnagle, J.H., Ticehurst, J., and Gerin, J.L.,** Hepatitis viruses, in *Diagnostic Procedures for Viral, Rickettsial, and Chlamydial Infections,* 6th ed., Schmidt, N.J. and Emmons, R.W., Eds., American Public Health Association, Washington, 1989, 957.

97. **Fraser, N.W. and Valya-Nagy, T.,** Viral, neuronal and immune factors which may influence herpes simplex virus (HSV) latency and reactivation, *Microbial Pathogen.,* 15, 83, 1993.

98. **Whitley, R.J.,** Herpes simplex viruses, in *Fields Virology,* 3rd ed., Vol. 2, Fields, B.N., Knipe, D.M., and Howley, P.M., Eds.-in-chief, Lippincott-Raven, Philadelphia, 1996, 2297.

99. **Rapp, F.,** Herpesviridae: herpes simplex virus types 1 and 2, in *Virology and Rickettsiology,* Vol. I, Part 2, Hsiung, G.-D. and Green, R.H., Eds., CRC Press, Boca Raton, 1978, 75.

100. **Marinesco, G.,** Recherches sur la pathologie de certaines encéphalomyélites à ultravirus, *Rev. Neurol.,* 1, 1, 1932.

101. **Balfour, H.H. and Heussner, R.C.,** *Herpes Diseases and Your Health,* University of Minnesota Press, Minneapolis, 1984, 38, 79, and 126.

102. **Nahmias, A.J., Keyserling, H.L., and Kerrick, G.M.,** Herpes simplex, in *Infectious Diseases of The Fetus and The Newborn Infant,* 2nd ed., Remington, J.S. and Klein, J.O., Eds., W.B. Saunders, Philadelphia, 1983, 636.

103. **Rozenberg, F.,** Virus herpes simplex, in *Virologie Moléculaire Médicale,* Seigneurin, J.-M. and Morand, P., Eds., Tec & Doc Lavoisier, Paris, and EM Inter, Cachan, 1997, 143.

104. **Nash, A.A. and Cambaropoulos, P.,** The immune response to herpes simplex virus, *Sem. Virol.,* 4, 181, 1993.

105. **Arvin, A.M.,** Varicella-zoster virus, in *Fields Virology,* 3rd ed., Vol. 2, Fields, B.N., Knipe, D.M., and Howley, P.M., Eds.-in-chief., Lippincott-Raven, Philadelphia, 1996, 2547.

106. **Grose, C.,** Variation of a theme by Fenner: the pathogenesis of chicken pox, *Pediatrics,* 68, 735, 1981.

107. **Hope-Simpson, R.E.,** The nature of herpes zoster. A long-term study and a new hypothesis, *Proc. Roy. Soc. Med.,* 58, 9, 1965.

108. **Straus, S.E.,** Clinical and biological differences between recurrent herpes simplex and varicella-zoster virus infections, *J. Am. Med. Assoc.,* 262, 3455, 1989.

109. **Mulder, W., Pol, J., Kimman, T., Kok, G., Priem, J., and Peeters, B.,** Glycoprotein D-negative pseudorabies virus can spread transneuronally via direct neuron-to-neuron transmission in its natural host, the pig, but not after additional inactivation of gE or gI, *J. Virol.,* 70, 2191, 1996.

110. **Calnek, B.W.,** Marek's disease and lymphoma, in *Oncogenic Herpesviruses,* Vol. I., Rapp, F., Ed., CRC Press, Boca Raton, 1980, 103.

111. **Hiyoshi, M., Tagawa, S., Takubo, T., Tanaka, K., Nakao, T., Higeno, Y., Tamura, K., Shimaoka, M., Fujii, A., Higashikata, M., Yasui, Y., Kim, T., Hiraoka, A., and Tatsumi, N.,** Evaluation of the AMPLICOR CMV test for the direct detection of cytomegalovirus in plasma specimens, *J. Clin. Microbiol.,* 35, 2692, 1997.

112. **Colimon, R. and Minjolle, S.,** Cytomégalovirus, in *Virologie Moléculaire Médicale,* Seigneurin, J.-M. and Morand, P, Eds., Tec & Doc Lavoisier, Paris, and EM Inter, Cachan, 1997, 169.

113. **Weller, T.H.,** The cytomegaloviruses: ubiquitous agents with protean clinical manifestations, II., *New England J. Med.,* 285, 267, 1971.

114. **Hengel, H., Brune, W., and Koszinowski, U.H.,** Immune evasion by cytomegalovirus - survival strategies of a highly adapted opportunist, *Trends Microbiol.,* 6, 190, 1998.

115. **Hill, A.B. and Masucci, M.G.,** Avoiding immunity and apoptosis: manipulation of the host environment by herpes simplex virus and Epstein-Barr virus, *Sem. Virol.,* 8, 361, 1998.

116. **Thorley-Lawson, D.A., Miyashita, E.M., and Khan, G.,** Epstein-Barr virus and the B cell: that's all it takes, *Trends Microbiol.,* 4, 204, 1996.

117. **Raab-Traub, N.,** Pathogenesis of Epstein-Barr virus and its associated malignancies, *Sem. Virol.,* 7, 315, 1996.

118. **Magrath, I., Jain, V., and Bhatia, K.,** Molecular epidemiology of Burkitt's lymphoma, in *The Epstein-Barr Virus and Associated Diseases,* Tursz, T., Pagano, J.S., Ablashi, D.V., de Thé, G., Lenoir, G., and Pearson, G.R., Eds., INSERM/John Libbey Eurotext, 1993, 377.

119. **Nash, A.A., Usherwood, E.J., and Stewart, J.P.,** Immunological features of murine gammaherpesvirus infection, *Sem. Virol.,* 7, 125, 1996.

120. **Gunton, P.E.,** Influenza A - clinical aspects and laboratory diagnosis, *Austr. Family Phys.,* 1, 343, 1972.

121. **Parker, C.E. and Gould, K.G.,** Influenza A virus - a model for viral antigen presentation to cytotoxic T lymphocytes, *Sem. Virol.,* 7, 61, 1996.

122. **Tashiro, M., and Rott, R.,** The role of proteolytic cleavage of viral glycoproteins in the pathogenesis of influenza virus infections, *Sem. Virol.,* 7, 237, 1996.

123. **Ahmed, R., Morrison, L.A., and Knipe, D.M.,** Persistence of viruses, in *Fields Virology,* 3rd ed., Vol. 1, Fields, B.N., Knipe, D.M., and Howley, P.M., Eds.-in-chief, Lippincott-Raven, Philadelphia, 1996, 219.

124. **Wright, T.C., Kurman, R.J., and Ferenczy, A.,** Precancerous lesions of the cervix, in *Blaustein's Pathology of the Female Genital Tract,* Kurman, E., Ed., Springer, New York, 1994, 229.

125. **Zur Hausen, H. and Schneider, A.,** The role of papillomaviruses in human anogenital cancer, in *The Papovaviridae,* Vol. 2, *The Papillomaviruses,* Salzman, N.P. and Howley, P.M., Eds., Plenum Press, New York, 1987, 245.

126. **Koss, L.G.,** Carcinogenesis in the uterine cervix and human papillomavirus infection, in *Papillomaviruses and Human Disease,* Syrjänen, K., Gissmann, L., and Koss, L., Eds., Springer, Berlin, 1982, 235.

127. **Jones, C.,** Cervical cancer: is herpes simplex virus type II a cofactor? *Clin. Microbiol. Rev.,* 8, 549, 1995.

128. **Pater, M.M., Mittal, R., and Pater, A.,** Role of steroid hormones in potentiating transformation of cervical cells by human papillomaviruses, *Trends Microbiol.,* 2, 229, 1994.

129. **Zur Hausen, H.,** Human papillomaviruses in the pathogenesis of anogenital cancer, *Virology,* 184, 9, 1991.

130. **Orth, G., Breitburd, F., Favre, M., and Croissant, O.,** Papovaviruses: possible role in human cancer, in *Origins of Human Cancer.* B. *Mechanisms of Carcinogenesis*, Hiatt, H.H., Watson, J.D., and Winsten, J.A., Eds., Cold Spring Harbor Laboratory Press, Cold Spring Harbor, 1977, 1043.

131. **Swan, D.C., Vernon, S.D., and Icenogle, J.P.,** Cellular proteins involved in papillomavirus-induced transformation, *Arch. Virol.,* 138, 105, 1994.

132. **Griffin, D.E. and Bellini, W.J.,** Measles virus, in *Fields Virology,* 3rd ed., Vol. 1, Fields, B.N., Knipe, D.M., and Howley, P.M., Eds.-in-chief, Lippincott-Raven, Philadelphia, 1996, 1267.

133. **Schneider-Schaulies, S., Schneider-Schaulies, J., Dunster, L.M., and Ter Meulen, V.,** Measles virus gene expression in neural cells, in *Measles Virus,* Termeulen, V. and Billeter, M.A., Eds., *Curr. Topics Microbiol. Immunol.,* 191, 101, 1995.

134. **Leinikki, P.,** Mumps, in *Principles and Practice of Clinical Virology,* 2nd ed., Zuckerman, A.J., Banatvala, J.E., and Pattison, J.R., Eds., John Wiley & Sons, Chichester, 1990, 375.

135. **Graham, B.S.,** Immunological determinants of disease caused by respiratory syncytial virus, *Trends Microbiol.,* 4, 290, 1996.

136. **Young, N.S.,** Parvoviruses, in *Fields Virology,* 3rd ed., Vol. 2, Fields, B.N., Knipe, D.M., and Howley, P.M., Eds.-in-chief, Lippincott-Raven, Philadelphia, 1996, 2199.

137. **Grulee, C. and Panos, T.C.,** Epidemic poliomyelitis in children, *Am. J. Dis. Children,* 75, 24, 1948.

138. **Swain, R.H.A. and Dodds, T.C.,** *Clinical Virology,* E. & S. Livingstone, Edinburgh, 1967, 101 and 167.

139. **Melnick, J.L.,** Enteroviruses: polioviruses, coxsackieviruses, echoviruses, and newer enteroviruses, in *Fields Virology,* 3rd ed., Vol. 1, Fields, B.N., Knipe, D.M., and Howley, P.M., Eds.-in-chief, Lippincott-Raven, Philadelphia, 1996, 655.

140. **Ginsberg, H.S.,** Picornaviruses, in *Microbiology,* 3rd ed., Davis, B.D., Dulbecco, R., Eisen, H.N., and Ginsberg, H.S., Eds., Harper & Row, Hagerstown, 1980, 1095.

141. **Bodian, D.,** Histopathological basis of clinical findings in poliomyelitis. *Am. J. Med.,* 6, 563, 1949.

142. **Smith, L.K. and Mabry, M.,** Poliomyelitis and the postpolio syndrome, in *Neurological Rehabilitation,* 3rd ed., Umphred, D.A., Ed., Mosby, St. Louis, 1995, 571.

143. **Bodian, D.,** Mechanisms of infection with poliovirus, in *Conference on Cellular Biology, Nucleic Acids, and Viruses,* v. St. Whitelock, O., Ed., N.Y. Acad. Sci., Special Publication No. V, 57, 1957.

144. **Huber, S.A.,** Animal models: immunological aspects, in *Viral Infections of the Heart,* Banatvala, J.E., Ed., Edward Arnold, London, 1993, 82.

145. **Douglas, R.G., Alford, B.R., and Couch, R.B.,** Atraumatical nasal biopsy for studies of respiratory virus infection on volunteers, *Antimicrob. Ag. Chemother.,* 18, 340, 1968.

146. **Straus, S.E.,** Viral hepatitis, in *Mechanisms of Microbial Disease,* 2nd ed. Schaechter, M.E., Medoff, G., and Eisenstein, B.I., Eds., Williams & Wilkins, Baltimore, 1993, 518.

147. **Swain, R.H.A. and Dodds, T.C.,** *Clinical Virology,* E. & S. Livingstone, Edinburgh, 1967, 101.

148. **Fenner, F.,** The pathogenesis of the acute exanthems. An interpretation based on experimental investigations with mousepox (infectious ectromelia of mice), *Lancet,* ii, 915, 1948.

149. **McFadden, G. and Graham, K.,** Modulation of cytokine networks by poxvirus: the myxoma virus model, *Sem. Virol.,* 5, 421, 1994.

150. **Yirrell, D.L., Norval, M., and Reid, H.W.,** Local experimental virus infections: comparative aspects of vaccinia virus, herpes simplex virus and human papillomavirus in man and orf virus of sheep, *FEMS Immunol. Med. Microbiol.,* 8, 1, 1994.

151. **Sharpe, A.H. and Fields, B.N.,** Pathogenesis of viral infections. Basic concepts derived from the reovirus model, *New England J. Med.,* 312, 486, 1985.

152. **Morrison, L.A. and Fields, B.N.,** Parallel mechanisms in neuropathogenesis of enteric virus infections, *J. Virol.,* 65, 2767, 1991.

153. **Bodkin, D.K., Nibert, M.L., and Fields, B.N.,** Proteolytic digestion of reovirus in the intestinal lumens of neonatal mice, *J. Virol.,* 63, 4676, 1989.

154. **Rubin, D.H., Kornstein, M.J., and Anderson, A.O.,** Reovirus serotype 1 intestinal infection: a novel replicative cycle with ileal disease, *J. Virol.,* 53, 391, 1985.

155. **Flamand, A., Gagner, J.-P., Morrison, L.A., and Fields, B.N.,** Penetration of the nervous systems of suckling mice by mammalian retroviruses, *J. Virol.,* 65, 123, 1991.

156. **Morrison, L.A., Fields, B.N., and Dermody, T.S.,** Prolonged replication in the mouse central nervous system of reoviruses isolated from persistently infected cell cultures, *J. Virol.,* 67, 3019, 1993.

157. **Nibert, M.L., Schiff, L.A., and Fields, B.N.,** Reoviruses and their replication, in *Fields Virology,* 3rd ed., Vol. 2, Fields, B.N., Knipe, D.M., and Howley, P.M., Eds.-in-chief, Lippincott-Raven, Philadelphia, 1996, 1557.

158. **Gallo, R.C.,** Leukemia, environmental factors, and viruses, in *Viruses and Environment,* Kurstak, E. and Maramorosch, K., Eds., Academic Press, New York, 1978, 43.

159. **Volk, W.A.,** *Essentials in Microbiology,* J.B. Lippincott, Philadelphia, 1978, 572.

160. **Huber, B.T., Hsu, P.-N., and Sutkowski, N.,** Virus-encoded superantigens, *Microbiol. Rev.,* 60, 473, 1996.

161. **Tambourin, P.E.,** Haematopoietic cells and murine viral leukaemogenesis, in *Stem Cells and Tissue Homeostasis,* 2nd Brit. Symp. Cell Biol., Lord, B.I., Potten, C.S., and Cole, R.J., Eds., Cambridge University Press, Cambridge, 1978, 259.

162. **González-Scarano, F., Nathanson, N., and Wong, P.K.Y.,** Retroviruses and the nervous system, in *The Retroviridae*, Vol. 4, Levy, J.A., Ed., Plenum Press, New York, 1992, 409.
163. **Hoover, E.A., Zeidner, N.S., Perigo, N.A, Quackenbush, N.L., Strobel, J.D., Hill, D.L., and Mullins, J.I.,** Feline leukemia virus-induced immunodeficiency syndrome in cats as a model for evaluation of antiretroviral therapy, *Intervirology,* 30 (suppl. 1), 12, 1989.
164. **Hause, W.R. and Olsen, R.G.,** Clinical aspects of feline leukemia diseases, in *Feline Leukemia,* Olsen, R.G., Ed., CRC Press, Boca Raton, 1981, 89.
165. **Hoover, E.A., Rojko, J.L., and Olsen, R.G.,** Pathogenesis of feline leukemia virus infection, in *Feline Leukemia,* Olsen, R.G., Ed., CRC Press, Boca Raton, 1981, 31.
166. **Yamaguchi, K., Kiyokawa, T., Futami, G., Ishii, T., and Takatsuki, K.,** Pathogenesis of adult T-cell leukemia from clinical pathologic features, in *Human Retrovirology: HTLV,* Blattner, W.A., Ed., Raven Press, New York, 1990, 163.
167. **Yoshida, M.,** HTLV-1 Tax: regulation of gene expression and disease, *Trends Microbiol.,* 1, 131, 1993.
168. **Höllsberg, P.,** Mechanisms of T-cell activation by human T-cell lymphotropic virus type I, *Microbiol. Mol. Biol. Rev.,* 63, 308, 1999.
169. **Narayan, O. and Clements, J.E.,** Lentiviruses, in *Virology,* 2nd ed., Vol. 2, Fields, B.N. and Knipe, D.M., Eds.-in-chief, Raven Press, New York, 1990, 1571.
170. **Blömer, U., Naldini, L., Kafri, T., Trono, D., Verma, I.M., and Gage, F.H.,** Highly efficient and sustained gene transfer in adult neurons with a lentivirus vector, *J. Virol.,* 71, 6641, 1997.
171. **Zink, M.C. and Narayan, O.,** Lentivirus-induced interferon inhibits maturation and proliferation of monocytes and restricts the replication of caprine arthritis-encephalitis virus, *J. Virol.,* 63, 2578, 1989.
172. **Wolthers, K.C. and Miedema, F.,** Telomeres and HIV-1 infection: in search of exhaustion, *Trends Microbiol.,* 6, 144, 1998.
173. **Bergeron, M.-G.,** Histoire naturelle du sida et l'avenir de l'épidémie et des traitements, in *VIH/Sida. Une Approche Multidisciplinaire,* Reidy, M. and Taggart, M.-E., Eds., Gaëtan Morin, Montréal, 1995, 567.
174. **Bolognesi, D.P.,** Human immunodeficiency virus vaccines, *Advan. Virus Res.,* 42, 103, 1993.
175. **Weissman, D. and Fauci, A.S.,** Role of dendritic cells in immunopathogenesis of human immunodeficiency virus infection, *Clin. Microbiol. Rev.,* 10, 358, 1997.
176. **Fauci, A.S.,** Basic immunology: the path to the delineation of the immunopathogenic mechanisms of HIV infection, *Trans. Assoc. Am. Physicians,* 101, clx, 1988.
177. **Fauci, A.S.,** The human immunodeficiency virus: infectivity and mechanisms of pathogenicity, *Science,* 239, 617,1988.
178. **Knight, S.C. and Patterson, S.,** Bone-marrow-derived dendritic cells, infection with human immunodeficiency virus, and immunopathology, *Annu. Rev Immunol.,* 15, 593, 1997.
179. **Pantaleo, G. and Fauci, A.S.,** Immunopathogenesis of HIV infection, *Annu. Rev. Microbiol.,* 50, 825, 1996.
180. **Rosenberg, Z.F. and Fauci, A.S.,** The immunopathogenesis of HIV infection, *Adv. Immunol.,* 47, 377, 1989.
181. **Kolson, D.L., Lavi, E., and González-Scarano, F.,** The effects of human immunodeficiency virus in the central nervous system, *Adv. Virus Res.,* 50, 1, 1998.
182. **Piot, P., Kapita, B.M., Ngugi, E.N., Mann, J.M., Colebunders, R., and Wabitsch, R.,** *Le SIDA en Afrique, Manuel du Praticien*, World Health Organization, Geneva, 1993, 36.
183. **Wigdahl, B. and Kunsch, C.,** Human immunodeficiency virus infection and neurologic dysfunction, *Prog. Med. Virol.,* 37, 1, 1990
184. **Bagasra, O., Lavi, E., Bobroski, L., Khalili, K., Pestaner, J.P., Tawadros, R., and Pomerantz, R.J.,** Cellular reservoirs of HIV-1 in the central nervous system of infected individuals: identification by the polymerase chain reaction and immunohistochemistry, *AIDS,* 10, 573, 1996.
185. **Atwood, W.J., Berger, J.R., Kaderman, R., Tornatore, C.S., and Major, E.O.,** Human immunodeficiency virus type 1 infection of the brain, *Clin. Microbiol. Rev.,* 6, 339, 1993.
186. **Levy, J.A.,** *HIV and the Pathogenesis of AIDS,* 2nd ed., ASM Press, Washington, 1994, 189, 191, and 198.
187. **Fishbein, D.B.,** Rabies in humans, in *The Natural History of Rabies,* 2nd ed., Baer, G.M., Ed., CRC Press, Boca Raton, 1991, 519.
188. **Robinson, P.A.,** Rabies virus, in *Textbook of Human Virology,* 2nd ed., Belshe, R.B., Ed., Mosby-Year Book, St. Louis, 1991, 517.
189. **Griffin, D.E. and Hardwick, J.M.,** Apoptosis in alphavirus encephalitis, *Sem. Virol.,* 8, 481, 1998.
190. **Banatvala, J.E.,** Laboratory investigations in the assessment of rubella during pregnancy, *Brit . Med. J.,* i, 561, 1968.
191. **Meyer, H.M., Parkman, P.D., and Hopps, H.E.,** The clinical application of laboratory diagnostic procedures for rubella and measles (rubeola), *Am. J. Clin. Pathol.,* 57, 803, 1972.
192. **Miller, L.K.,** Insect viruses, in *Fields Virology,* 3rd ed., Vol. 1, Fields, B.N., Knipe, D.M., and Howley, P.M., Eds.-in-chief, Lippincott-Raven, Philadelphia, 1996, 533.
193. **Steinhaus, E.A.,** *Principles of Insect Pathology,* McGraw-Hill, New York, 1949, 168, 172, 462, 494, and 505.

194. **Volkman, L.E.,** The 64K envelope protein of budded *Autographa californica* nuclear polyhedrosis virus, in *The Molecular Biology of Baculoviruses,* Doerfler, W. and Böhm, P., Eds., *Curr. Topics Microbiol. Immunol.,* 131, 1986, 103.

195. **Fraser, M.J.,** Transposon-mediated mutagenesis of baculoviruses: transposon shuttling and implications for speciation, *Ann. Entomol. Soc. Am.,* 79, 773, 1986.

196. **Tweeten, K.A., Bulla, L.A., and Consigli, R.A.,** Applied and molecular aspects of insect granulosis viruses, *Microbiol. Rev.,* 45, 379, 1981.

197. **Paillot, A.,** *L'Infection chez les Insectes. Immunité et Symbiose,* Imp. G. Pâtissier, Trévoux, 1933, 98 and 99.

198. **Crawford, A.M. and Sheehan, C.,** Replication of *Oryctes* baculovirus in cell culture: viral morphogenesis, infectivity and protein synthesis, *J. Gen. Virol.,* 66, 529, 1985.

199. **Matthews, R.E.F.,** *Plant Virology,* 3rd ed., Academic Press, San Diego, 1991.

200. **Gray, S.M.,** Plant virus proteins involved in natural vector transmission, *Trends Microbiol.,* 4, 259, 1996.

201. **Gray, S.M. and Banerjee, N.,** Mechanisms of arthropod transmission of plant and animal viruses, *Microbiol. Mol. Biol. Rev.,* 63, 128, 1999.

202. **Leisner, S.M., Turgeon, R., and Howell, S.H.,** Long-distance movement of cauliflower mosaic virus in infected turnip plants, *Mol. Plant Microbe Interact.,* 5, 41, 1992.

203. **Leisner, S.M. and Howell, S.H.,** Long-distance movement of viruses in plants, *Trends Microbiol.,* 1, 314, 1993.

204. **Eglinton, G. and Hamilton, R.J.,** Leaf epicuticular waxes, *Science,* 156, 1322, 1967.

205. **Chay, C.A., Gunasinge, U.B., Dinesh-Kumar, S.P., Miller, W.A., and Gray, S.M.,** Aphid transmission and systemic plant infection determinants of barley yellow dwarf luteovirus-PAV are contained in the coat protein readthrough domain and 17-kDa protein, respectively, *Virology,* 219, 57, 1996.

206. **Garner, R.J.,** *The Grafter's Handbook,* 5th ed., Faber and Faber, London, 1988, 133, 134, 139, 144, and 187.

207. **Samuel, G.,** The movement of tobacco mosaic virus within the plant, *Ann. Appl. Biol.,* 21, 90, 1934.

208. **Esau, K.,** An anatomist's view of virus diseases, *Am. J. Bot.,* 43, 739, 1956.

209. **McLean, B.G., Waigmann, E., Citowsky, V., and Zambryski, P.,** Cell-to-cell movement of plant viruses, *Trends Microbiol.,* 1, 105, 1993.

210. **Stevens, W.A.,** *Virology of Flowering Plants,* Blackie & Sons, Glasgow, 1983, 28, 29, 30, 35, 36, and 95.

211. **Esau, K.,** Pathological changes in the anatomy of leaves of the sugar beet, *Beta vulgaris,* affected by curly top, *Phytopathology,* 23, 679, 1933.

212. **Hatta, T., and Francki, R.I.B.,** Anatomy of virus-induced galls on leaves of sugarcane infected with Fiji disease virus and the cellular distribution of virus particles, *Physiol. Plant Pathol.,* 9, 321, 1976.

213. **Strobel, G.A. and Mathre, D.E.,** *Outlines of Plant Pathology,* Van Nostrand Reinhold, New York, 1970, 282.

214. **Hatta, T., Bullivant, S., and Matthews, R.E.F.,** Fine structure of vesicles induced in chloroplasts of Chinese cabbage leaves by infection with turnip mosaic virus, *J. Gen. Virol.,* 20, 37, 1973.

215. **Goldin, M.I.,** The physical action of viruses on the plant cell, in *Viruses of Plants,* Proc. Internat. Conf. Plant Viruses, Beemster, A.B.R. and Dijkstra, J., Eds., North-Holland Publishing Company, Amsterdam, 1966, 158.

216. **Cornuet, P.,** *Éléments de Virologie Végétale,* INRA, Paris, 1987, 17.

INDEX

Viral vernacular names and disease designations derived from these are listed under the corresponding scientific names; for example, *Papillomaviridae* stands for papillomaviruses and papillomas. Commonly abbreviated names are listed by their abbreviations followed by full names.

A

Abortion, 28, 109
ADCC, antibody-dependent cell-mediated cytotoxicity, 58, 65, 179, 233
Adenoviridae, 5, 8, 9, 15, 19, 20, 33, 34, 68, 73
African swine fever, 8
Age dependence, 8, 75, 77, 140
Aggregation, 54, 56
AIDS, acquired immunodeficiency syndrome, 1, 122, 176-194; see HIV
neurologic, 182, 189, 190
Aleutian mink disease, 14, 141, 233
Alfalfa mosaic virus, 211
Alphaherpesvirinae, 102-117
Alpharetrovirus, 166
Alphavirus, 197
Anemia, 15, 165, 166
Antibody, 17, 44, 50, 51, 53, 55, 57-59, 233
Antibody production, longterm, 60
Antigen-antibody complex, 55, 185
Antigen recognition phase, 53
Aphids, 201, 213
Aphthovirus, 143
Apoptosis, 1, 2, 7, 8, 36-38, 73, 124, 155, 187, 197, 233
Arabis mosaic virus, 224, 233
Arboviruses, 1, 7, 19, 86, 87, 197
Arenaviridae, 6, 19, 74-77
Arteriviridae, 3, 5
Arthralgia, 86, 197
Arthritis, 141, 165, 199
Ascoviridae, 5, 201
Astroviridae, 5, 20
Aujeszky's disease, 102, 116, 233
Australia antigen, 91
Autoimmunity, 71, 72, 150, 151, 165, 168, 174, 190, 233
Aviadenovirus, 73
Avian leukosis, 28, 166
Avian type C retroviruses, 166
Avihepadnavirus, 91

B

B cell, 8, 47, 50, 51, 60, 62, 71, 122-126, 163, 233
B19 parvovirus, 141
Baculoviridae, 5, 201, 202, 204-209
Barley stripe mosaic virus, 224, 233
Betaherpesvirinae, 102, 118-121
Betaretrovirus, 167
Birnaviridae, 6, 201
Blood-CNF barrier, 189
Bornaviridae, 6, 8
Bovine diarrhea, 86
Bowenoid lesion, 132
Bracovirus, 202
Bronchiolitis (RSV), 140
Bunyaviridae, 1, 6, 7, 78, 79, 201, 211
Burkitt's lymphoma, 8, 122, 125, 193, 233
Bursa of Fabricius, 4, 45, 117

C

Caliciviridae, 5, 20, 80
CaMV, cauliflower mosaic virus, 217, 233
Cancer, carcinoma, 165-167, 194
anogenital, 130, 134
cervical, 130, 132
liver, 8, 90-92, 97-100, 154
mammary, 167
nasopharyngeal, 8, 122, 125
Canine distemper, 137
Caprine arthritis-encephalitis, 174, 175
Carcinogenesis, 97 130-136
Carcinoma *in situ*, 131, 233
Carrier state, 91, 96, 98, 99, 101, 233
Caudovirales, 3, 5
CD4 cells, 9, 50, 62, 66, 67, 72, 88, 89, 111, 126, 173, 174, 176, 177, 180-183, 188, 193, 194
CD8 cells, 62, 66, 67, 70, 72, 88, 89, 118, 121, 126, 173, 177, 186
Cell fusion, 8, 31
Chemotaxis, 46, 53, 61
Chickenpox, see Varicella-zoster
Chloroplast, 212, 221-225, 233
Chordopoxvirinae, 155
Circulative virus, 211, 213, 233
Cirrhosis, 90-92, 154
Classification of viruses, 3-6
CMV, *Cytomegalovirus*, 9, 19, 20, 118-121
CNS infection, 24-26

Common cold, 81, 143, 153
Complement, 3, 44, 45, 53, 54, 57, 58, 64, 89, 201, 233
Condyloma, 130
Conjunctivitis, 75
Contagious pustulous dermatitis, 155
Coronaviridae, 3, 5, 19, 20, 81-83
Cowpea mosaic virus, 211
Cowpox, 8
Coxsackievirus, 8, 20, 143, 150
Creutzfeldt-Jakob disease, 19, 235
Curly-top virus, 218, 233
Cypovirus, 201
Cytokines, 9, 36, 44, 46-52, 62-65, 67, 68, 72, 84, 86, 88, 89, 95, 121, 126, 138, 140, 155, 157, 169, 172-174, 190, 192-194, 234

D

Deltaretrovirus, 172
Deltavirus, 1, 6, 7, 101
Dendritic cell, 64, 158, 178, 180, 183, 184, 234
Dengue, 20, 86, 88, 89
Densovirinae, 141
Dependovirus, 141
Diarrhea, 159

E

Ebola virus, 20, 84
ECHO virus, 20, 143
Ectodesmata, 220, 234
Ectromelia, 155, 156
Enamovirus, 211
Encephalitis, 8, 14, 15, 25, 78, 87, 105, 107, 116, 137, 138, 165, 197
Encephalomyelitis, 81, 137, 138
Enterovirus, 12, 19, 20, 143, 153
Entomopoxvirinae, 155, 201, 234
Entry sites of viruses, 9
Episome, 7, 16, 104, 110-122, 124, 125, 130, 131, 234
Epstein-Barr virus, 8, 9, 12, 19, 20, 122-125, 193, 194, 234
Equine encephalitis, 25
Equine infectious anemia, 15
Erythroblastosis, 166
Erythrovirus, 141
Evasion of immune surveillance, 7, 9,